THE PRIME VIBRATION

A THEORY OF EVERYTHING EMERGING FROM LOVE

MICHAEL SMITH

The Prime Vibration
Copyright © 2021 by Michael Smith

All rights reserved. No part of this publication may be reproduced, distributed, or transmitted in any form or by any means, including photocopying, recording, or other electronic or mechanical methods, without the prior written permission of the author, except in the case of brief quotations embodied in critical reviews and certain other non-commercial uses permitted by copyright law.

Tellwell Talent
www.tellwell.ca

ISBN
978-0-2288-5052-6 (Paperback)

For everyone questioning the nature of reality
and why love seems to be the answer
at the heart of it all

CONTENTS

Introduction vii

PART ONE
The Universal Thoughtform of Love Becoming

1. **Searching for Meaning and Truth at the Heart of It All** 3
   The Measurement Problem of Perceiving the Truth 4
   The Scientific Aversion to Nonphysical Truth 5
   The Vibrational Nature of Expressing Our Truth 8
   The Collective Truth of Emergence 10
   Case Study: COVID-19 as a Catalyst of Truth 12

2. **Imagining the Simplest Theory of Everything** 14
   Un-Learning to See Simply 14
   Describing Reality in the Simplest Terms 16
   A Simple Picture of Reality Emerges 19
   Reality Unfolding from 2D Projection of 3D Double-Helix 28
   Redefining Reality and Who We Are 31

3. **Discovering a Conscious Universe Emerging from Love** 34
   Consciousness: The Geometry of Love Knowing Itself 35
   The 12D Crystalline Memory of Consciousness 38
   The Fractal Emergence of Love 44
   The Emergence of Love at The Universal Scale 46
   The Emergence of Love at The Galactic Scale 63
   The Emergence of Love at The Scale of Life 73
   The Emergence of Love at The Quantum Scale 88
   The Emergence of Love at The Fundamental Scale 130

PART TWO
Becoming More Resonant with Love

4. **Perceiving Love Through your Intuitive Senses** 137
   Your Extra-Sensory Abilities...139
   To Feel: Following Your Emotional Guidance 140
   To See: Looking Beneath The Surface.......................................143
   To Hear: Listening To What Isn't Said......................................146
   To Know: Trusting Your Inner Guidance.................................148

5. **Exploring Love Through the Base-12 Numerology Cycle**...151
   The Twelve Themes of the Base-12 Numerology Cycle 152
   Navigating Love along the Path of Resonance 158
   The Holy Grail of Love and the Incarnation Cycle..................166
   Love In The Balance 173
   The Numerology of Nomenclature.. 175

6. **Mastering Love Through the Universal Laws of Vibration**...181
   The Six Universal Laws of Vibration 182
   Polarity: Finding Clarity In Contrast 184
   Wavelength: Experiencing Full Cycles of Potential 187
   Frequency: Exploring Specific Vibrational Themes.................190
   Amplitude: The Amplifying Power of Emotions......................192
   Resonance: Manifesting with the Law of Attraction 194
   Octave: Mastering Higher Frequencies 198
   Mastery of Play 202
   The Collective Mastery of Soul Groups.................................... 204
   Mastering the Spirituality of Science 205

Notes 215

INTRODUCTION

We are all searching for essentially the same thing: a happy, abundant and meaningful life. For most of us that likely includes good health, financial security, the freedom to pursue our passions, making a positive difference in the world and sharing love. But most important of these is love for without love everything else seems incomplete and less fulfilling. Still, we also know that life has its share of challenges, set-backs and detours along the way which test our ability to sustain love. For the most part, we take these trials and tribulations in stride, dusting ourselves off and soldiering forward with optimism and hope for a more loving tomorrow.

In recent years, however, the barrage of global crises and unrest has been intense: from the 9/11 terrorist attacks of 2001 and ensuing wars in Afghanistan and Iraq, the Enron scandal and stock market crash of 2002, SARS outbreak of 2003, Hurricane Katrina in 2005, the global financial crisis of 2008, Haiti earthquake of 2010, Ebola outbreak of 2014 and, of course, the global COVID-19 pandemic of 2020. Many of us have experienced major upheavals in our personal lives too during this emotional roller coaster ride.

It can be difficult to remain positive and loving in such trying times as these, searching for direction and meaning within all the chaos and confusion. We all want to believe that love is the greater purpose of everything we experience and who we are but it can be hard to discern love within hardship. After all, love is more of a feeling or emotion than something objective and tangible we can prove as a fundamental truth.

But what if we *could* prove that love is the creative force behind it all? What if we could understand how love really works, why dramatic events unfold and how to navigate those events in the most loving and beneficial way? The great news is that now we can. This book deciphers the hidden vibrational pattern through which love emerges, why love seeks out challenge and contrast and how everything that unfolds in our lives we vibrationally attract. And remarkably, this love story is expressed in the most fundamental language of all – numbers and geometry.

Going back to the basics of what mathematics says is most fundamental – the prime numbers – and the most efficient way to express the whole in terms of its parts – the base-12 number cycle, an elegant vibrational pattern is revealed. Taking the shape of a double-helix waveform emerging from a central frequency of 6, this surprisingly simple pattern seems to explain everything we physically perceive as well as the very nature of consciousness itself – of a neutral point of perspective gaining awareness of itself through creative expressions of itself.

The geometry of the twelve frequencies within the base-12 prime pattern also precisely match their vibrational meanings from numerology, the ancient intuitive science of the energetic qualities of numbers. As the 6 frequency represents the vibrational theme of *love*, we suddenly realize a profound "Theory of Everything" emerging from love itself – where love is the source and purpose of it all and the prime vibration the dynamic through which love manifests.

The Prime Vibration puts you back in control of your life with greater clarity, connection and purpose than ever before. You will learn how to tune in to the frequency of love through your intuitive senses, how to explore love using the base-12 numerology cycle and how to master love through the laws of vibration. The ultimate good news story, this book decodes and demystifies the intelligent and unified field of consciousness through which love explores and enlightens itself. Knowing we all embody that same profound purpose reminds us of the loving creators we are and that the truth is always at the heart of it all.

Love's Universal Signature Hiding In Plain Sight

Sometimes it takes a radically new way of looking at things to see our way forward. A mechanical engineer and statistician for many years, I was your classic "numbers guy" and a defiant skeptic of anything I couldn't see, touch or prove to myself – including spirituality. I certainly wasn't interested in far-flung theories of the universe or existential questions about the nature of reality, consciousness and love. Not for this guy thank you very much – that is, until 2014.

That was the pivotal year in which I experienced a sudden spiritual awakening and mediumship ability to connect with the consciousness of those who had passed. Confronted with the realization that our consciousness lives on after physical death, my whole notion of reality was fatally flawed and obsolete. This propelled me into metaphysics and (non-meta) physics in search of answers and, unbeknownst to me at the time, was a synchronistic fit with my other sudden obsessions with *numerology*: the ancient intuitive science of the energetic qualities of numbers, *prime numbers*: the fundamental building blocks of mathematics, and the *base-12 number system*: a highly efficient way of expressing the whole in terms of its parts and the pattern preferred by nature.

I had no idea what the connection could possibly be between these three radically different ways of looking at numbers – or with my spiritual path – until I happened to combine them together in a particular way. When I graphed the four positions within the base-12 cycle where prime numbers can occur (i.e. at 1, 5, 7 and 11), a beautiful double-helix vibrational pattern emerged. Not only did this pattern look remarkably like the shape of the DNA molecule – the fundamental blueprint of life itself – but also the geometry of each of the twelve number positions within the waveform precisely matched the ancient number meanings of numerology intuited thousands of years ago. This led me to adapt traditional base-10 Pythagorean numerology to base-12, enabling far more accurate and insightful numerology readings than possible with the base-10 approach. My first book, *"Base-12 Numerology: Discover Your Life Path Through Nature's Most Powerful Number"* (Llewellyn Worldwide), explores the powerful base-12 numerology system and provides a step-by-step guide to completing your own personal numerology reading.

Seeing the base-12 prime pattern in action through numerology, I could certainly appreciate how well it worked as an energetic blueprint for translating quantitative number frequencies into qualitative themes we experience. However, I was still perplexed *why* it worked and whether the prime pattern extended beyond numerology. The engineer in me just had to know and couldn't let it go (apparently mistaken for stubbornness by spouses of engineers).

The more I investigated, the more I realized that the prime waveform seemed to explain practically *everything* we see at all scales of nature: from the Big Bang and curved spacetime to dark energy and a seemingly expanding universe, from black holes and gravity to dark matter and the shape of galaxies, from atoms and subatomic structure to the properties of the particles and forces, from DNA and the genetic code to photosynthesis and the golden ratio of organic growth, and from red blood cells and the respiratory system to neurons and the human brain.

It then dawned on me that the perpetual infinity-shaped pattern of the prime waveform also seemed to reflect the conceptual geometry of consciousness – of a central neutral point of reference gaining awareness of itself through creative expressions of itself. If consciousness *and* everything which manifests into physical reality do indeed follow the same prime pattern then this would explain why numerology works too: physical reality itself emerges from consciousness according to the prime waveform and numerology simply describes the vibrational geometry or energetic personality of each number frequency within that waveform.

This led me to the most profound realization of all: if everything is a manifestation of consciousness emerging from the 6 frequency of *love* at the heart of the prime waveform then this means *everything emerges from love*! This is the beautiful truth I believe is at the heart of it all. We are all expressions of love connected through love on a journey of discovery together. So I guess you could say this is the ultimate love story – a love story we always felt were true in our hearts and can finally know for certain.

How This Book Is Organized

The Prime Vibration is organized into six chapters (of course). We begin in Chapter 1 with the challenge we face; of coming to grips with a jaded world seemingly spinning out of control and a scientific world view equally confusing, fragmented and spiritually averse. We then wipe the slate clean in Chapter 2, rethinking the very nature of reality in the simplest terms possible – as cycles of prime numbers in base-12. It is through this radical new lens that a vibrational picture of reality suddenly comes to life as a universal harmonic pattern of consciousness emerging from love. The many amazing ways this fundamental pattern agrees with the physical world around us and with our spiritual reality within are examined in Chapter 3.

The second half of the book puts the prime vibration into practice, providing simple and effective strategies for harnessing the power of love in our daily lives. In Chapter 4 we learn how to fine-tune our connection to love through our intuitive senses. In Chapter 5 we learn how to apply the base-12 numerology cycle as our road map for exploring love in a proactive and empowered way. Finally, in Chapter 6 we learn how to master love by following the six universal laws of vibration through which love manifests and achieves balance: *Polarity*: Finding clarity in contrast, *Wavelength*: Experiencing full cycles of potential, *Frequency*: Exploring specific vibrational themes, *Amplitude*: Amplifying power of emotions, *Resonance*: Manifesting with the Law of Attraction, *Octave*: Mastering higher frequencies.

PART ONE

The Universal Thoughtform of Love Becoming

1

Searching for Meaning and Truth at the Heart of It All

It's a paradox of our times: the more technologically advanced society becomes, the more vulnerable and separate we tend to feel. Perhaps this is an inevitable trade-off that comes with our modern age of convenience – with consumables, goods and services easily procured comes a certain loss of self-reliance and with a digitally connected world less personal interaction. Still, the loss we feel on an individual and collective level seems to go much deeper than this. It's as if we're missing a core part of ourselves and drifting away from what connects us to each other. And in the particularly turbulent times of late we feel even more disconnected and adrift.

Our current scientific world view doesn't offer much solace either. Despite incredible strides made in understanding how the physical universe works from the cosmically large to the quantumly tiny, little focus has been given to the non-physical aspects of reality. And by non-physical I mean consciousness and love; the two qualities which seem to define us at the most fundamental level. Granted, physical reality is inherently more measurable than something as elusive and ethereal as consciousness or love. Nevertheless, deciphering and understanding consciousness and love remain the most important mysteries of all in making sense of our place in the big picture – of embracing our truth.

The Measurement Problem of Perceiving the Truth

To perceive or measure the non-physical aspects of reality would presumably need non-physical means of doing so. Fortunately, we already possess this technology, each and every one of us. We come fully equipped at birth with extra-sensory abilities of perception which are natural extensions of our physical senses though simply underutilized and rusty from lack of use. And, like any measurement device, the more precise the gauge the finer the measurements we can make. So it stands to reason that the more finely tuned and higher in frequency our own consciousness the higher levels of consciousness and information we should be able to perceive beyond ourselves. In Chapter 4 we will explore how we can fine-tune our intuitive senses to attain and sustain those higher frequencies.

Still, we are each clouded by various biases, insecurities and judgmental criteria through which we see and judge the outside world – how we measure reality. And much of this measurement bias stems from our evolutionary instinct for survival, of whether we perceive something as safe or not. Because of this self-preservation reflex we tend to react to situations as either good or bad personally while ignoring the bigger picture collectively. How we judge ourselves and others follows similarly narrow pass/fail criteria; fat or thin, smart or ignorant, worthy or underserving, confident or fearful, friendly or distant, boxers or briefs, and so on. So when we search for reasons why things happen the way they do, especially unpleasant things, we tend to measure them with the same judgmental and critical yardstick.

The educational system and religious institutions can also perpetuate this pass/fail mentality, either knowingly or inadvertently. In school, students are expected to achieve certain grades based on rather arbitrary and sporadic testing throughout the year, effectively rewarding conformance while forcing students to learn at the same pace. Likewise in church, the congregation is expected to follow a set doctrine and sequence of teachings while obeying a strict code of conduct including when to listen, read, sing, sit, stand or kneel. Those who don't conform to the prescribed beliefs or methods are then judged negatively. This unfortunately gives school and religion a bad rap, especially to those who don't like being told how to learn or what to believe. School essentially becomes a limited pass/fail version of education and religion a pass/fail variant of spirituality. This is not to say that standardized education and religious formality is a bad thing by any means, as many find great comfort and reassurance in being guided through educational and spiritual teaching in a structured and consistent way. It's just when that structure becomes overly restrictive and the guidance becomes mandatory with little room for personal discovery that the line of personal freedom is crossed.

As we will discuss often in the chapters ahead, this "measurement problem" of what we allow ourselves to perceive and how freely we allow each other to do so is perhaps the greatest barrier to love that humanity imposes on itself. This is therefore our greatest opportunity too.

The Scientific Aversion to Nonphysical Truth

The scientific community likewise tends to pass judgement on what it deems acceptable to explore. This determines which areas of research receive funding, which research papers scientific journals will publish and which candidates become tenured professors in academia. Those who venture too far outside the mainstream and along nonphysical tangents such as consciousness, psychic phenomena or afterlife communication generally do so on their own with little support. I find this ironic as the goal of science is the open exploration of all possible avenues of inquiry provided sound scientific method and validation is followed. After all, yesterday's radical and outlandish idea may become tomorrow's relativity theory or quantum mechanics. So many of our greatest discoveries were at first discounted as nonsense.

One such idea that went from vilified to valid is wave-particle duality. This is the observed phenomenon in which energy and matter can behave both as waves and particles depending on conditions: they can diffract and interfere with each other like waves but can also act as point-like particles when measured or observed. All objects appear to possess this inherent split personality though only detectable at the small scale of quantum particles such as electrons and photons of light. Although science is able to mathematically describe this odd behaviour in terms of probabilities and wave equations, what it really represents remains a mystery – especially why the conscious act of simply observing an experiment causes the vibrational wave-like behavior to seemingly collapse into a fixed particle state. This "measurement problem" of quantum physics is one of the most glaring pieces of evidence we have that consciousness and the physical reality we perceive are intimately connected and warrants much further study.[1]

Fortunately, there's a growing number of organizations, research collaborations and individuals who are bravely exploring the non-physical realm despite the materialist tunnel vision of the scientific status quo. Several such pioneering organizations include the Institute of Noetic Sciences (IONS), the Foundational Questions Institute (FQXi), the John Templeton Foundation and the John E. Fetzer Memorial Trust, all of which I happened upon in my intuitive journey and who unknowingly helped bolster my resolve to write this book. Their vision and mission statements say it all:

- The Institute of Noetic Sciences, founded by Apollo 14 astronaut Edgar Mitchell, conducts research into frontier topics in consciousness *"to reveal the interconnected nature of reality through scientific exploration and personal discovery"*.[2]
- The Foundational Questions Institute, directed by MIT physicist Max Tegmark, sponsors grants, essay contests and conferences *"to catalyze, support, and disseminate research on questions at the foundations of physics and cosmology, particularly new frontiers and innovative ideas integral to a deep understanding of reality but unlikely to be supported by conventional funding sources"*.[3]
- The John Templeton Foundation sponsors and funds activities which advance its vision of *"infinite scientific and spiritual*

> *progress, in which all people aspire to and attain a deeper understanding of the universe and their place in it"*.[4]
- The John E. Fetzer Memorial Trust through its Fetzer Franklin Fund sponsors and supports efforts to *"ensure the integrity of the founder's spiritual vision, by seeking to advance integrated, relational views of reality through exploring scientific frontiers and universal spiritual practices"*.[5]

It is this unbiased search for the truth these organizations embrace and promote which attracts those who are willing to think outside the box and explore fresh new ideas, just like the leading edge pioneers who founded them and those who champion them today.

The Institute of Noetic Sciences was founded by the Apollo astronaut Edgar Mitchell after a deeply spiritual experience he had when viewing Earth from space and its chief scientist Dean Radin is world-renowned in consciousness research. The Foundational Questions Institute (FQXi) was co-founded and is directed by MIT physicist Max Tegmark, a top physicist and influential proponent of the universe being mathematical in nature. The John Templeton Foundation was established by the late Sir John Templeton, a pioneer in financial investment and philanthropy and an open-minded advocate for the discovery of spiritual truth and the ultimate nature of reality. Lastly, the John E. Fetzer Memorial Trust is the legacy of the late John E. Fetzer who is best known as a Michigan entrepreneur, icon of broadcasting and owner of the Detroit Tigers baseball team in the 1960's. He is less well known for his life-long exploration of diverse spiritual philosophies and metaphysical systems, the continued pursuit for which he established his trust.[6] Actually, it was my happening upon the annual FQXi essay contest, of which both the Fetzer and Templeton organizations have been regular sponsors, that I first came to learn of their extensive support for spiritually inclusive science and the stories of their inspirational founders.

The following timeless quote by John Fetzer captures perfectly what I have also come to believe and intuitively know to be true: "Love is the core energy that rules everything. It's the force field of creation." This same belief was held dear by John Templeton as reflected in the title of the book which, poignantly in the last days of his life, he asked to be written on his behalf, "Is Ultimate Reality Unlimited Love?".[7]

I can think of no better role models than John Fetzer and John Templeton for courageously pursuing one's truth wherever it may lead while achieving both material and spiritual abundance in a values-driven way. Although I hadn't the privilege of meeting either of these two visionaries I somehow feel we are kindred spirits writing this book together in our exploration of love, both scientifically and spiritually.

The Vibrational Nature of Expressing Our Truth

Like wave-particle duality, another undeniable truth of science is that everything is energy in vibrational form. From light to sound to atoms, it is through vibration that energy and matter are expressed. This book you are holding and the chair you are sitting on are comprised of countless atoms ordered together in structured ways. Although mostly empty space, these objects only seem solid because of the clouds of electrons surrounding each atom's nucleus and the electromagnetic force keeping them apart. It's not surprising then that the vibrational nature of reality is expressed in so many vibrational phenomena such as electromagnetism, gravitational waves, the ocean tides, seismic undulations of the Earth's crust, the rhythmic beat of the heart, the fertility cycle, brain wave activity and the circadian rhythm of our daily waking cycle.

What you may not realize is just how vibrationally structured our daily lives are too. Every experience we encounter and relationship we have follows a cycle as does every idea, discussion, project, course, career or life time. Indeed, every experience of any kind represents a vibrational cycle of beginning, engagement and ending through which we encounter life and gain deeper understanding. The same applies to teams, companies, organizations, countries and entire societies. All of these collaborations follow a fundamental cycle of learning, growth and development of some sort or another. But how well that cycle progresses for each depends on how the collective group and all of its constituents are in synch with each other and with the same shared objectives. In other words, how vibrationally aligned they are to their truth.

We can find many fine examples of groups, organizations and companies which vibrationally embody such unity and purpose. And they're usually easy to spot, not only through the lasting success and positive influence they achieve but also how well they exemplify the six fundamental laws of vibration:

Polarity: To find clarity in contrast, manage opportunity and adversity in a balanced and objective way and to learn what works and what doesn't

Wavelength: To recognize that everything is part of a natural process and cycle and that honoring each step of that cycle without taking shortcuts is the key to consistent success

Frequency: To perform each task and activity with focus, attention and care, seeking mastery of the individual skills required and utilizing personal talents to best advantage

Amplitude: To be engaged and energized with high commitment, enthusiasm and passion

Resonance: To manifest their vision into reality by resonating with it through the Law of Attraction: being clear about what they want to achieve (frequency), energizing that goal by preparing for it, expecting it and getting excited about it (amplitude), and then allowing the best outcome to emerge without restricting it's form (wavelength).

Octave: To continuously improve and achieve higher levels of excellence while staying true to yourself and your core values.

One such organization I feel exemplifies the six laws of vibration is Apple Inc. Rated in 2020 by Fortune Magazine as the world's most admired company for thirteen years running and by Forbes as the most valuable brand, Apple is leading edge in practically every business metric that matters.[8, 9] These include innovation (*octave*), people management and quality of management (*amplitude*), use of corporate assets and social responsibility (*wavelength*), financial soundness and long-term investment value (*resonance*), quality of products and services (*frequency*) and global competitiveness (*polarity*).

Known for its relentless innovation in realizing ever higher levels of potential in its products and its brand, Apple has remained relevant and a dominant player in the ultra-competitive personal computer and technology market. Even the numerology of its name "Apple" resonates with this innovative theme of *new beginnings (1) through new beginnings (1) of potential (0)* (i.e. "Apple" = 17735 = 1x12 + 11 = 1 + 11 = 1x12 + 0 = 10/1). So too does the Life Path (birth date) of visionary co-founder, Steve Jobs (February 24, 1955 = 2 + 20/2 + 11611/17/8 = 2 + 2 + 8 = 1x12 + 0 = 10/1). As we will explore in Chapter 5, numerology offers an accurate universal language for understanding our vibrational reality – personally, professionally and globally.

In Chapter 6 we will come to appreciate just how universal these six vibrational principles are as enablers of greater understanding and mastery of our lives. Once we allow ourselves to think, feel and act with vibrational freedom, we resonate with our true authentic nature as vibrational loving beings. And once we do that, we resonate together in powerful ways and create lasting harmony. Our collective love emerges.

The Collective Truth of Emergence

This brings us to perhaps the most important key to unlocking our human potential: together we become more. This phenomenon is called *emergence* in science, whereby complex behaviour and order spontaneously self-organizes or *emerges* from collections of simpler elements or rules.[10] New and sophisticated things that groups of individuals can do together they cannot do alone; emergence is the collective truth that harmony creates.

We see emergence throughout nature: in the way flocks of birds and schools of fish can choreograph their movements as a single entity, how billions of neurons work together in the brain, how colonies of ants or termites work together to create complex nests, how clouds of interstellar gas coalesce into self-sustaining stars and how billions of stars emerge into cohesive galaxies and those galaxies into distinct clusters.

We see this in human behavior too in the way groups of individuals naturally gravitate together and take on a collective personality and culture: close friends tending to dress and act alike, players on sports teams able to anticipate each other's movements and work as one and entire societies, countries and historical periods having a collective culture or zeitgeist of their own. And as history has dramatically shown, such periods can embody a wide range of values from enlightened to oppressive or anywhere in between.

Nevertheless, a prerequisite for emergence to occur in any of its forms is for the individuals within a collective to be like-minded or otherwise similar in nature. In cities, for example, multiple smaller communities emerge from those individuals who share the same ethnicity, culture or lifestyle. These cities-within-cities are a healthy emergent phenomenon of diverse urban society.

There are also those individuals who do not conform to any particular group of similarly aligned individuals. These outliers stray from the collective momentum of the consensus. Still, sometimes they can have such an influential presence that, over time, the momentum shifts towards them and what collectively emerges conforms to them. The influence of these charismatic and/or forceful few can be benevolent or otherwise. Civil rights and peace movements are the product of this type of championed shift as are dictatorships and terrorist cells. However, as I hope to validate in this book, we are all created of love and therefore most naturally resonate with benevolent and peaceful emergence, not oppression. And the more we understand and assert our inherent goodness the fewer will be swayed by misguided catalysts of non-love.

Case Study: COVID-19 as a Catalyst of Truth

Every so often in the history of humanity a global event occurs which is so impactful it snaps everyone's attention onto a single focal point. These include world wars, natural disasters and pandemics. The most recent of these, of course, is the COVID-19 pandemic which began in late 2019 and will likely impact society for many years to come.

Moreso than any event in recent memory, the pandemic has highlighted the polarity of unity versus division, cooperation versus blame and science versus capitalism. Although appearing entirely negative on the surface and a big step back for society, this polarity has actually vibrationally amplified the contrast of what we value most and is thus a powerful catalyst to make more empowered and compassionate choices for the world we share.

From the heightened polarity caused by the pandemic and the social distancing implemented to reduce its spread, new and creative ways of living, working and learning from home have emerged. Although this has presented major challenges in all aspects of society as we adjust to our new reality, it has also afforded greater work life balance, a more economical and local lifestyle and reduced traffic congestion and pollution. This has translated into a greater consideration for the personal space of others and our direct impact on the environment whilst, at the same time, encouraging kindness and compassion in helping each other get through this together. Trying times such as these profoundly test our capacity for good and our ability to see beyond our self-imposed and divisive distinctions of nationality, race and economic interests.

In vibrational terms, the pandemic is helping us gain a more objective and neutral perspective through a broader understanding of our polarized potential. And, as with any cataclysmic event that shakes the world order, such polarity stimulates both fear and solidarity and thus tests our strength and resolve in the face of change. We will return to this idea often in this book of achieving personal and collective resilience through vibrational contrast.

In this opening chapter we have examined our search for the truth at the heart of which is a fundamental yearning for love. We considered how this search has only been heightened by the dramatic times we have experienced of late, catalytic events which have tested our emotional and spiritual resolve to see love in all things. We also discussed the various obstacles we face in perceiving the full truth of our reality including the measurement problem of trying to measure nonphysical reality through strictly physical means, our instinctive bias of judging things as good versus bad and our scientific bias of neglecting the nonphysical in favor of what we can only physically see, touch and/or measure.

We then described the nature of truth itself; how natural phenomena and human experience are both fundamentally vibrational and follow the same basic vibrational laws, notably emergence. It is this quality of emergence – of complex order naturally emerging from simple fundamental elements and properties – that sets the stage for Chapter 2 in which we rethink reality in the simplest terms possible. From that simplest "Theory of Everything" we discover how consciousness and everything else may indeed emerge from the vibration of love, profoundly illuminating what is ultimately true.

2

Imagining the Simplest Theory of Everything

As mentioned in the Introduction, sometimes we need to go back to basics when we reach an impasse in making progress or solving something. This could be a baffling homework problem at school, a challenging technical problem at work or a frustrating relationship issue at home. Whatever the issue, we can reach the point where we have tried using our current approach from every conceivable angle but to no avail and simply need to start again from scratch. Sometimes we reach that decision to hit the reset button gradually on our own while other times it happens suddenly, like my spiritual awakening in 2014.

Un-Learning to See Simply

Although I didn't realize it at the time, there were plenty of signs that my old materialistic world view and skeptical outlook were no longer working for me – if indeed they ever did. I went through a series of increasingly frustrating and short-lived job changes in my manufacturing career leading up to 2014 and a worsening lack of fulfillment with these roles. Much of my training and career progression up until that point had been in process improvement and operations management, focusing on increasing productivity and reducing cost. To put it nicely,

I was more or less an "efficiency expert"; basically a corporate hatchet man tasked with the unpopular job of minimizing labor and squeezing the most production out of the fewest resources. Ironically, once I completed the dirty work of some of these reorganization assignments, I too was sent packing. Touché, Monsieur Smith!

As the years went on, I grew more and more disillusioned with the admittedly cold and calculating work I did and the unfulfilling trajectory I was on. As my level of stress climbed, my overall health suffered too – to the point of dangerously high blood pressure and several severe back-to-back bouts of pneumonia. I was physically, emotionally and spiritually a mess.

Then in the Fall of 2014 my spiritual reset button was pushed in a dramatic way. I began experiencing lucid dreams of family members who had passed as well as clairvoyant images and premonitions of things that would happen days later. I immediately had a knowing – which actually felt more like a remembering – that our souls live on after physical death and that we are far more spiritually powerful and connected to each other's consciousness than we allow ourselves to believe. So in one fell swoop I went from engineer skeptic to spiritual seeker.

The other knowing that hit me like a two-by-four squarely on the noggin was that I couldn't continue doing the work I had been doing. Within days of these psychic experiences beginning, I began to feel physically ill and exhausted before going to work each morning and it took all my strength and intestinal fortitude to drive to work and make it through the day. With the unwavering support and love of my wonderful wife through all of this, I ended up quitting my job shortly thereafter. This not only snapped me out of my midlife funk and dis-ease but launched me on a personal quest to understand what was happening to me and why.

Over the next year or so I read numerous books on metaphysics and spirituality, Googled a googol of new age videos and took every class I could find including meditation, psychic development, mediumship, channeling and energy healing. Although my metaphysical hopscotch from one area of study to another must have seemed completely erratic to friends and family, I actually felt I was being intuitively led through a specific sequence of learning and discovery I needed to follow. Synchronicity seemed to be paving the way and I was just following its road map. I must confess it took a lot for me to get out of my own head and just allow and trust these subtle feelings and inner guidance after decades of trying to rationalize my way through everything. I was essentially un-engineering myself to allow me to see what was essentially true.

In order to see the simplest and most fundamental truths, we need to learn how to see simply and that often involves unlearning the many biased assumptions with which we over-complicate reality. The more biases we burden ourselves with, the more layers we need to shed. For me, it felt like a spiritual full monty. Sorry for that mental image.

Describing Reality in the Simplest Terms

My mediumship ability developed rapidly during this time and I began offering readings professionally in 2015. This is also when my focus was drawn to numerology. Although already mathematically inclined as an engineer and statistician, I became absolutely obsessed with numerology and the ancient idea that numbers have vibrational properties including anything that can be described by numbers – which is basically everything. I also became fascinated with prime numbers, the base-12 number system and physics. Again, this wasn't really out of intellectual curiousity but rather a strong intuitive pull to explore these areas for some reason.

Numerology

Numerology interprets whole numbers (and letters as numbered positions in the alphabet) as vibrational frequencies, where any number or word can be *reduced* down to its simplest single-digit frequency or note within the first octave of the number cycle. This reduction is done by adding together the individual digits of the original number to get a single digit. The number 35, for example, reduces down to a single-digit frequency of 3 + 5 = 8. In math jargon, 8 is called the "digital root" of 35 when counting in the base-10 number system.

Each number frequency also has a specific meaning or *theme* with which it resonates, where the final digit (8 in our example) is the main *outer energy* while the multiple digits which add together to get that final number (3 + 5) are the *inner influences* contributing to it. Here, our original number 35 means *Manifestation* (8) *through the Catalyst* (3) *of Change* (5). This is the vibrational signature of the *Change Agent*; one who manifests results (8) by influencing (3) change (5).

Numerology also works in reverse. By subtracting an *inner influence* number from the *outer energy* number, we can reveal what the outer frequency changes to. Using our same example above of 3 + 5 = 8, simply moving the 3 from the left side of the equation to the right gives 5 = 8 − 3 and shows how the 8 energy changes to a 5 when the influence of the 3 is removed. In other words, an outcome (8) without an intentional cause (3) is merely unintentional change (5).

Prime Numbers

Where numerology describes the fundamental unique frequencies from which all numbers emerge, prime numbers are the fundamental unique numbers from which all integers emerge. A prime number is any integer greater than 1 that's only divisible by 1 and itself. Thus, the numbers 2, 3, 5, 7, 11 and 13 are the first handful of primes. Also, all possible whole numbers are either primes or products of primes. This is why prime numbers are often called the building blocks of mathematics.

As we saw above, numerology uses addition (or subtraction) to determine how numbers vibrationally combine (or cancel out). Prime numbers instead use multiplication (or division) to determine how numbers mathematically combine (or reduce to their prime factors). That is, every non-prime integer is the product of two or more prime numbers as factors.

The non-prime integer 30, for instance, is the product of prime factors 2, 3 and 5 (i.e. 2 x 3 x 5 = 30). Conversely, we can work backwards to find the prime factors from which any non-prime integer is derived. This is done by dividing the non-prime number by prime factors until you get a prime that can't be divided any further (e.g. 30 / prime 5 = 6 and 6 / prime 3 = prime 2).

Base-12

The base-12 number system represents an alternative way of expressing numbers in terms of cycles. Unlike the base-10 or *decimal* system we use today for counting and calculating most things and which counts in cycles of ten from 0 to 9, the base-12 or *duodecimal* system counts in cycles of twelve from 0 to 11, where 10 and 11 are treated as single-digit numbers (underlined here to distinguish them from their double-digit counterparts 10 and 11). This is the familiar "clock cycle" we still use to track the hours of the day, a legacy from earlier times of the well recognized usefulness of base-12.

Although there are simpler number base systems out there, notably the base-2 or *binary* system used by computers which counts in cycles of just two (from 0 to 1), base-12 is highly efficient and versatile for its still compact cycle. This is because base-12 offers more ways of evenly dividing itself than most other alternate bases. For example, where base-10 can only divide 10 evenly two ways (in halves: 2 x 5 or fifths: 5 x 2), base-12 can divide 12 four ways (in halves: 2 x 6, thirds: 3 x 4, quarters: 4 x 3 or sixths: 6 x 2). This superior factorability for its size makes base-12 a natural choice as a fundamental number system. As nature tends to operate in the most efficient way, it's not surprising that the base-12 pattern occurs so often from lunar cycles to snowflakes and subatomic particles. Not to mention boxes of doughnuts and beer.

Now, something I didn't appreciate at first was that all three of these ways of looking at numbers – vibrationally through numerology, mathematically through prime numbers and structurally through base-12 – are arguably the most efficient and fundamental ways of doing so. It must have been for this same reason I was also being intuitively drawn to study physics as it is through physics we seek to describe the fundamental nature of reality. Perhaps I was being led to un-learn physics to reveal a simpler truth about reality just as I had to un-learn my worldview to see things more simply.

Still, an even bigger question remained: how might these supposedly *simplest* concepts of numerology, prime numbers and base-12 translate into the *simplest* theory of reality? The exciting answer to this question and synchronistic story of how it came to light we explore next.

A Simple Picture of Reality Emerges

As part of my unabashed free fall down the rabbit hole of all things number related, I came across a famous unsolved math problem involving prime numbers called the Riemann Hypothesis.[11] Without going into the gory details, this notoriously unsolved problem in mathematics is concerned with proving or disproving whether the positive "zero values" of a certain complex function all lie on the same vertical line with a real value of 0.5. As this function and its zeros are related to the prime numbers, I thought perhaps I would discover something interesting by poking this hibernating beast. Right away I could tell most of it was way over my head but gave it a look nevertheless.

While unsuccessfully trying to decipher this formidable problem from a base-12 angle I did, however, notice a repeating pattern hidden within the prime numbers themselves. I found that the prime numbers, in addition to 1 but excluding 2 and 3, all seem to fall on the same four positions of the base-12 circle when viewed as a cycle. These four positions are 1, 5, 7 and <u>11</u> as shown below (again, ten and eleven underlined to denote them as single digits in base-12). Although this base-12 pattern of all primes falling on these four positions is nothing new, I don't believe it had been previously examined as a vibrational cycle.

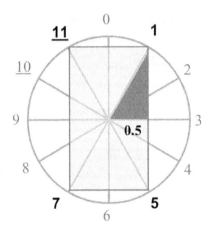

The Base-12 Number Cycle

Another thing that caught my eye was how all four of the prime positions of 1, 5, 7 and 11 as coordinates around the base-12 circle have a horizontal coordinate of plus or minus 0.5. So if the base-12 circle were viewed as the unit circle (i.e. a circle of radius 1) in the complex plane such that the vertical axis is the imaginary number line i and the horizontal axis the real number line x, then all four prime positions as $x + iy$ complex number coordinates have an absolute real value of 0.5. A further curious property of the base-12 cycle is that the square of any number from those four prime positions will always fall on prime position 1 and therefore always have a positive real value of 0.5. As this is the same property the Riemann Hypothesis seeks to validate I felt I was on to something promising.

Alas, my limited math background wasn't up to pursuing this enticing clue further into the mathematical abyss of the Riemann Hypothesis. Still, perhaps it may help some smarty-pants in eventually wrestling this problem to the ground some day. If it does, all I ask is that they buy me a nice bottle of wine from their million dollar Millennium Prize up for grabs by the Clay Mathematics Institute.[12]

Anyway, when I graphed rotations of those four prime positions as sine waves with amplitudes equal to their values, the prime positions appeared as whole number intersection points through the real number axis as depicted below. This suggested that the prime numbers may indeed be vibrational on a fundamental level; a conclusion I had no problem making as I already considered numbers as having energetic qualities through my study of numerology.

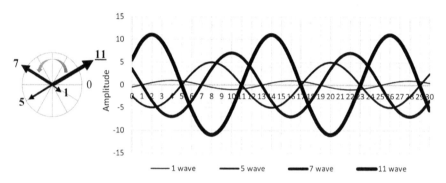

Vibrational Behavior of Base-12 Prime Positions 1, 5, 7 and <u>11</u>

In the above graph you may notice that the four sine waves intersect the horizontal axis as two pairs of waves intersecting at the same locations. That is, the 1 and 7 waves always share the same intersection points, such as at 1, 7, 13, 19… etc., as do the 5 and <u>11</u> waves at 5, 11, 17, 23…, etc. This is because the two sine waves within each pair are 180-degrees opposite each other on the base-12 circle. This causes each pair of waves to have opposite amplitudes, but otherwise remain in phase with each other, allowing us to subtract each pair into a combined wave through destructive interference without affecting their shared intersection points.

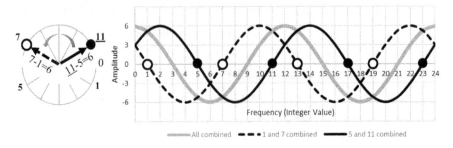

Vibrational Behavior of Base-12 Prime Positions 1+7 and 5+<u>11</u>

This produces the dual wave pattern illustrated above, where the dashed black line represents the 7 - 1 combined wave and the solid black line the <u>11</u> - 5 combined wave. This elegant vibrational pattern of two overlapping waves, shifted sideways from each other by four base-12 positions but otherwise of identical amplitude 6 and wavelength 12, repeats every twelve integers. It immediately struck me how remarkably similar it is to the two-dimensional side view of the double-helix DNA molecule. Also, wherever these two waves intersect the horizontal number line (emphasized by black and white dots) are where the prime numbers can occur, again, along with 1 but excluding 2 and 3. I'll explain what's special about 1, 2 and 3 shortly.

When I combined those two overlapping prime waves together they produced the single pure cosine wave shown in grey. This grey line therefore seemed to represent the fundamental vibrational chord of the primes, the path along which each number's vibration achieves balance individually and over which all twelve numbers achieve harmony as an octave. I realized this dual waveform could indeed be the fundamental structure of the prime numbers when numbers themselves are considered vibrational frequencies rather than just lifeless quantities. As that's exactly what numerology is, I was confident I had found the link between numerology, the primes and base-12 for which I was searching.

Still, numerology ventures further by giving specific qualities or personalities to each number frequency within the cycle and to double-digit "master numbers" such as 11, 22 and 33. These number meanings have stood the test of time, remaining unchanged for thousands of years. So, I wondered how the traditional base-10 number meanings might fit into this new base-12 pattern I had uncovered, if at all. When I applied the traditional base-10 numerology definitions to the base-12 prime waveform, I was shocked to see that the geometry of the waveform at each number matched its number meaning precisely, as depicted below.

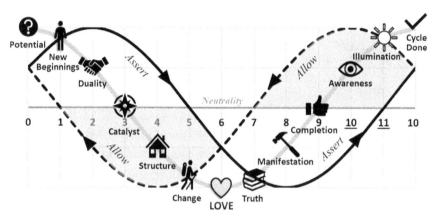

The Base-12 Numerology Cycle

For example, the 0 of *potential* at the very start of the cycle appears as a singularity where the two polarities (solid and dashed black waves) converge with maximum unrealized potential (amplitude of the grey wave above the neutral horizontal axis). The 1 of *new beginnings and independence* possesses the singular geometry of positive polarity only and the first encounter with the physical (dashed black wave intersecting the neutral axis). The 2 of *duality and polarity* appears as the 2:1 geometry of twice the positive polarity than negative. Likewise, the 3 of the *catalyst*, that which causes change without changing itself, reflects a geometry of neutral balance achieved (grey wave intersecting the neutral axis) through the maximum expression of positive and negative polarity in equal measure. And, of course, the all-important 6 of *love* where the two polarities intersect as one in a pure state of neutrality and maximum potential achieved (amplitude of the grey wave below the reality axis).

Accurate geometric representations of the rest of the twelve numbers are similarly mirrored by the base-12 prime vibration. This not only seemed to validate the ancient numerological intuition of numbers as vibrations with specific meanings from an objective mathematical perspective, but also suggested to me that the true language of numerology and perhaps mathematics itself has indeed been base-12 all along.

I was so intrigued by this possibility that I proceeded to adapt traditional base-10 numerology to the base-12 number system. I was already impressed by the uncanny ability of base-10 numerology to reveal much about a person's life path theme of experiences (derived from the vibration of their birth date) and how they express themselves (based upon the vibration of their birth name). However, re-tuning numerology to base-12 gave even more accurate and insightful results when applied to individuals whose life stories I knew well, such as my own, family, friends and famous personalities.

What's Special about 1, 2 and 3: Redefining the Prime Numbers in Base-12

Since the number 12 consists of the factors 2 and 3 (i.e. 2 x 2 x 3 = 12), the numbers 2 and 3 can be considered the underlying structure of the base-12 cycle itself, not of the primes generated by that cycle. This is why primes can never occur at positions 2 or 3 for subsequent base-12 cycles nor at any product which includes one or both of those numbers, such as 4 (2 x 2), 6 (2 x 3), 8 (2 x 2 x 2), 9 (3 x 3), 10 (2 x 5) or 10 (2 x 2 x 3). This leaves positions 1, 5, 7 and 11 as the only possible positions for primes to occur. Not only does this suggest that 2 and 3 should be *excluded* from the set of prime numbers, but also that 1 should be *included* since position 1 in the base-12 cycle is necessarily one of the four base-12 positions critical to the primes, contrary to the established definition of the prime numbers that excludes 1 but includes 2 and 3.

Indeed, once we treat numbers as frequencies within vibrational cycles of twelve rather than just tick marks along a linear number line, the traditional definition of primes suddenly becomes obsolete and even misleading. So instead of *"any integer greater than 1 that is only divisible by 1 and itself"* I suggest a more meaningful definition of the prime numbers would be *"1, 5, 7 or 11 or any multiple of 12 above which is only divisible by 1 and itself"*. That is, the first series of primes expressed in base-12 are 1, 5, 7, 11, 1x12+1=11 (13 in base-10), 1x12+5=15 (17 in base-10), 1x12+7=17 (19 in base-10)… and so on. This highlights how the prime numbers are not just the numerical building blocks of all numbers but also their fundamental musical notes.

Balanced Geometry of the Base-12 Cycle

Now, the cyclical nature of prime numbers isn't unique to base-12. It occurs in the base-10 number system too. However, instead of 2 and 3 being the underlying structure of the cycle as in base-12, it is 2 and 5 in base-10 (since 2 x 5 = 10). As such, primes can never occur at positions 2 or 5 in the base-10 cycle nor at any product that includes one or both of those numbers, such as 4 (2 x 2), 6 (2 x 3), 8 (2 x 2 x 2) or 10 (2 x 5). This leaves positions 1, 3, 7 and 9 as the only possible positions for primes to occur in base-10. Once again, we see the number 1 as a necessary *cyclical* factor of the primes. Still, this is a well known property of primes in base-10, so again nothing new.

The important difference between base-10 and base-12 when viewed as cycles, however, is that the four prime positions in base-12 possess a balanced rotational geometry whereas in base-10 they do not, as illustrated below. This I believe is a key property of prime numbers overlooked in mainstream mathematics and the underlying reason why the base-12 number system is the logical vibrational choice of nature through which to express itself.

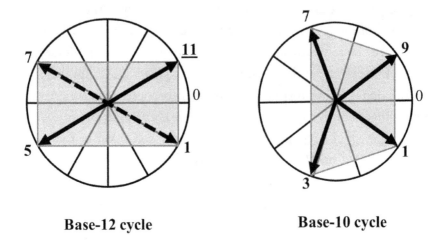

Base-12 cycle   **Base-10 cycle**

Geometry of the Prime Positions in Base-12 versus Base-10

In base-12, the four prime positions create a symmetric rectangular geometry which, when expressed as vibrational sine waves, reduces to the two balanced waves of the base-12 prime vibration. Again, this is because the 1 and 7 positions are opposite each other on the base-12 circle, as are 5 and 11, such that each pair subtracts to a single wave of amplitude 6 (7 − 1 = 6, 11 − 5 = 6). Note the 2:1 ratio reflected by the rectangular geometry of the base-12 prime positions, of four positions wide by two positions tall. This fundamental 2:1 relationship occurs throughout nature and mathematics and, as we will discover in Chapter 3, is the reason why our reality operates through the duality of polarity and contrast.

In base-10, there's no such balanced symmetry or combining of opposite frequencies. Instead, the four positions remain as four separate sine waves of varying amplitude of 1, 3, 7 and 9 when graphed. The resulting combined waveform becomes a vibrationally discordant mess with no apparent pattern or simplicity, which would explain why finding any pattern to the primes has proven so elusive in base-10. This further highlights how the arbitrary choice of base-10 as the worldwide standard for mathematics actually obscures the underlying base-12 structure of numbers.

What About Non-prime "Gaps" Generated from the 1, 5, 7 and <u>11</u> Positions?

One final question about the primes we need to address before moving on is why the primes have unpredictable gaps between adjacent primes, gaps that tend to get further apart as the prime numbers get larger.

According to the conventional definition of prime numbers, these gaps are any integers divisible by another integer other than 1 and are therefore disqualified as primes. The first example of this in base-10 is 25 (2 x 12 + 1 = 21 in base-12) as it is divisible by 5 and the next is 35 (2 x 12 + <u>11</u> = 2<u>11</u> in base-12) being divisible by 5 or 7. But this too can be explained once we treat the prime numbers as unique vibrational intersection points between the base-12 prime vibration and the horizontal real number line. Every prime "candidate" generated from the base-12 positions of 1, 5, 7 or <u>11</u> represents its own unique frequency with its own corresponding wavelength. As such, we can picture each prime number as a separate sine wave of a specific wavelength superimposed over the base-12 prime waveform of wavelength 12, along with the sine waves of all previous primes.

As the requirement for a prime number is that it cannot be evenly divisible by any number other than 1, this is equivalent to a prime candidate sine wave not being evenly divisible by the wavelength of any previous prime – or in terms of frequency, not being a harmonic multiple of any smaller prime. Since 25 (21 in base-12) is a harmonic multiple of the lower prime frequency of 5, it fails this uniqueness test and is therefore not a prime. This also explains why the gap between consecutive primes tends to get larger as the prime numbers get larger; there are more and more previous primes as potential harmonics to create those gaps.

The base-12 prime vibration itself provides a quick visual way to check for such ineligible prime candidates: if a prime candidate (i.e. an intersection point of the base-12 prime waveform with the integer axis) occurs anywhere a previous prime's sine wave also intersects, it's not a prime. That's because the prime candidate would have to be some vibrational multiple of a lower prime for it to "fit" and have a shared intersection point.

This reinterpretation of prime factors as base-12 harmonics of other primes could possibly impact the field of cryptography, the encrypting of confidential information, as it typically uses calculations with extremely large prime numbers due to the difficulty in determining their prime factors. If it turns out that those factors can more easily be found through a vibrational base-12 approach, then banks and credit card companies may some day be very upset at me. Still, I would expect the courtesy of yet another bottle of wine for this heads-up, especially as my credit cards may suddenly stop working upon my next visit to the wine store.

Reality Unfolding from 2D Projection of 3D Double-Helix

Now, something I didn't realize at first was that my two-dimensional plot of the base-12 prime waveform only showed part of the picture. That's because I had only plotted the vertical or sine coordinate of each prime position on the base-12 circle but not its horizontal cosine coordinate. Once I did, the full 3D double-helix shape of the prime waveform came into view as shown below. Also shown, shaded grey, is the "front view" projection of that 3D shape as a 2D surface.

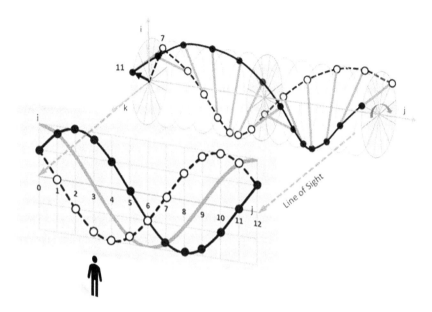

Base-12 Prime Waveform as the 2D Surface of a 3D Double Helix

A property of the 2D surface of a 3D shape is that it can be holographically encoded with all the information about its 3D form.[13] This means that when that encoded surface is projected back into 3D all the properties of its original form are retained and at all fractals of scale. If the base-12 prime waveform is indeed holographic then any physical 3D manifestation which emerges, such as an atom, DNA molecule, galaxy or universe, will be a full and faithful vibrational copy of the original – including having a heart-centered focus of love. This suggests that the reality we perceive as three spatial dimensions plus time may simply be a holographic projection of the base-12 prime waveform as the fundamental blueprint of potential from which everything emerges.

This idea of physical reality emerging holographically from a non-local enfolded geometry is not a new one. Building upon earlier work of Louis de Broglie from the 1920's, the theoretical physicist David Bohm proposed in 1952 a quantum theory called the de Broglie-Bohm pilot wave theory which posits the existence of a hidden pilot wave which guides the path of any particle.[14] Bohm further expanded this theory to suggest that two levels of reality exist; an unfolded or *explicate* order of physical reality and an enfolded or *implicate* order from which physicality emerges in a holographic way. This included the notions of space, time, physical matter and perhaps even consciousness all being emergent phenomena from a deeper and more fundamental structure. Also, like a hologram, the full implicate structure is inherent in every unfolded aspect of itself. In this way, everything is part of and interconnected with the undivided whole. The holographic principle was further formalized in 1993 by the Dutch theoretical physicist Gerard 't Hooft.[15]

The prime waveform seems to align conceptually well with Bohm's theory; the 3D double helix representing the pilot wave of fundamental *implicate* order, its 2D boundary the surface on which all properties of the prime waveform are encoded and 4D space-time the *explicate* holographic projection from that surface. It was my chance encounter with the John E. Fetzer Memorial Trust in which I first learned of David Bohm's work as he had served as a scholar in residence at the Fetzer Institute during the latter part of his career.

Recent advances in theoretical physics further support this idea of 4D space-time emerging holographically from the 2D surface of a 3D quantum system. Perhaps the most compelling development is what is called the AdS/CFT correspondence of string theory proposed in 1997 by the Argentinian-American physicist Juan Maldacena.[16] This correspondence (which links "anti-de Sitter spaces" and "conformal field theories") shows how spacetime including gravity could emerge like a hologram from the boundary of a lower-dimensional quantum particle model that excludes gravity. The ongoing challenge with this theory, however, is that the type of universe it describes is a finite bounded one with a negatively ("anti-") curved saddle-like space that contracts inward upon itself – not the positively curved space of an infinitely expanding universe like the one we seem to live in. As we will see in the next chapter, the prime waveform suggests a way out of this cosmic quandary: our universe may be negatively curved after all with its twisting rotation mistaken for linear expansion.

We will also uncover a compelling explanation for the mysterious wave-particle duality so fundamental to physics: everything exists as undefined potentials in wave form and are only perceived as fixed particles where the prime waveform intersects the neutral reality axis – at prime number positions 1, 5, 7 and 11. It is therefore at these four positions where the fundamental or "prime" particles of matter can occur (spoiler alert: the up and down quarks and the electron and tau neutrinos). And just as all possible natural numbers are either prime numbers themselves or products of primes, all possible forms of matter are either prime particles or combinations thereof.

The geometry of the prime waveform suggests a further important property. As the waveform represents just a single cycle of the base-12 prime cycle, this standing wave would manifest into physical reality as a figure-8 experiential loop. Consequently, with each additional cycle of experience returning through the same central point of position 6 but taking another of infinitely many possible paths, the prime waveform would take the form of a torus – the characteristic doughnut shape of atoms, magnetic fields, spiral galaxies and perhaps the multiverse. Perhaps this is why I like doughnuts so much. Hmm?... more research needed.

Redefining Reality and Who We Are

Now, I appreciate that everything I have suggested so far is, like a multiverse-sized doughnut, a lot to digest: that numbers are energetic vibrations with distinct personalities, that ancient civilizations intuitively knew this and applied it through the mystical practice of numerology, that the prime numbers have a hidden base-12 vibrational pattern which has evaded detection for thousands of years, that base-10 – the standard number system used worldwide – is dysfunctional and should be dropped in favour of base-12 and, the doozy of them all, that reality itself is a holographic projection of consciousness in the form of a simple base-12 vibration emerging from the 6 frequency of *love*. And all this from a retired engineer who talks to dead people and barters in bottles of wine as payment for math tips. In earlier times I would likely have been burned at the stake as a heretic. I probably wouldn't fare too well in the mathematics or physics department of any self-respecting university today either unless, of course, I shared my wine.

All joking aside, I was convinced that the base-12 prime waveform I had uncovered was the key to understanding our reality and our place within it, and that more would be revealed to me. Its elegant double-helix shape also deeply resonated with me, mirroring the DNA blueprint of life, the ancient symbol for infinity and the polarized nature of duality itself. The possibility that such a simple geometric shape could be the truth beneath it all is a captivating thought.

From the simplest conceptual way of describing reality – of prime numbers as vibrations – a base-12 pattern emerges which not only agrees with modern mathematics and ancient numerology but also, as we will explore in Chapter 3, the nature of consciousness and so much of our physical reality too. From the evolution of the universe, the behaviour of galaxies and black holes, the life cycle of stars, the growth pattern of plants, the shape of the DNA molecule and the structure of the atom and its subatomic particles, the base-12 prime pattern appears to fit remarkably well as the underlying truth.

Still, the most profound realization of all is that this seemingly universal blueprint of reality and consciousness emerges from the 6 frequency of *love* at the very heart of the waveform. It is this beautiful idea – that love is at the heart of it all – which fundamentally resonates with all that we see, all that we feel and all that we are. This reveals us to be creative expressions of Love/Source/God experiencing, exploring and knowing itself. We are therefore all connected and part of the same unity of All That Is. Our separateness is merely an illusion of our physical form, but an invaluable one, in order that we may freely explore the infinitely many ways of finding our way back to the truth. And this necessarily means experiencing contrast, polarity and challenge to help us navigate ever more adventurous and illuminating paths home to love.

I also believe we are only presented with challenges and opportunities we are consciously ready to handle; that everything we face we are capable of navigating and understanding. Thus, the high level of global and personal contrast we have been encountering of late reflects just how ready and able we are for greater learning and growth. Love is showing us how capable we are of elevated levels of compassion and harmony despite appearances to the contrary. We simply need to believe and embrace that in body, mind and soul. So whenever we experience physical, emotional or spiritual imbalance or dis-ease, we are actually on the threshold of discovering a new way back to love and are being provided contrast to help illuminate the way. And with each unique rediscovery of love we expand the collective wisdom of the whole of which we are part.

In the second half of the book we will discuss practical and powerful ways of using the prime vibration as a road map for living vibrationally empowered lives and up to our loving potential. But before we do, let's explore the many amazing ways the base-12 prime waveform seems to explain what we perceive to be true consciously, physically and spiritually.

3

Discovering a Conscious Universe Emerging from Love

In the previous chapter we considered a radically new way of looking at reality; by imagining it in the simplest terms possible. Still, this was really just a conceptual exercise – a thought experiment in which we suspended everything we think we know about reality and reimagined it from the ground up. Sure, it produced a pretty vibrational picture that happens to match numerology and looks a lot like DNA. And, yes, the 6 frequency of *love* being at the heart of this idealistic picture of reality and consciousness is something we so want to be true. But maybe I'm reading way too much into this twisted little pattern. Perhaps that spiritual two-by-four I took to the head in 2014 did some permanent damage. Seems far too simple to be true, doesn't it?

That's what I thought too until I started applying it to the world around us. From the cosmos to the quantum world and the scale of life in between, the base-12 prime vibration seems to explain it all and more. I know that's a pretty audacious claim to make but I think you will be just as surprised and intrigued once you weigh the evidence for yourself.

In this chapter we will take a tour through nature with the prime vibration as our guide. I will touch on major sights along the way and show how, amazingly, everything we perceive can be rationally and simply explained by our beautiful little theory of love. As what we perceive starts with consciousness, that's where our tour starts too.

Consciousness: The Geometry of Love Knowing Itself

When we think of consciousness (pun intended), we consider the way our minds work in perceiving and understanding the world around us. As such, what we perceive is always relative to ourselves as an individual "Me" having those thoughts; how we relate to our environment and how it relates to us. In other words, our self-awareness.

This gives us a working definition of consciousness as *an awareness of self through experiences relative to the self.* This also implies a further defining property of consciousness: that its central point of awareness, of the self, is a neutral zero-point of reference. This means that all experiences relative to the self must maintain that overall neutrality too so as to return to the same neutral state in order to be understood. In other words, experience must occur in an equally polarized way relative to its origin. We could call this self-referential property of consciousness the *conservation of identity* – that the neutral sense of self can neither be created nor destroyed, only reinformed.

Based on the above definition of consciousness, the prime waveform would seem to be the perfect geometric description – *of love (as the neutral intersection point at position 6) becoming aware of itself through equally polarized expressions of itself.* And, again, with each additional unique figure-8 cycle of experience returning through the same central point of awareness, the overall flow of consciousness would take the form of a torus where each cycle represents a vertical "slice". To help picture this, shown below is a stripped down version of our prime waveform representing one such figure-8 cycle or slice of experience. This will serve as our model of consciousness.

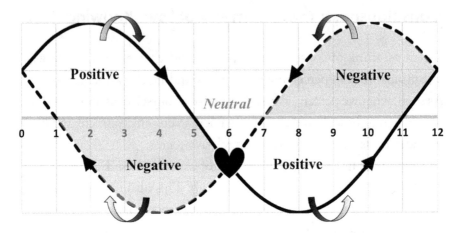

Flow of Consciousness

Let's walk through this diagram, beginning with what the horizontal and vertical axes represent. The horizontal axis is the real number line spanning the twelve positions of the base-12 cycle, the various number frequencies or "themes" that can be experienced in any given cycle. Because real numbers represent actual physical quantities grounded in reality, the real number line is neutral by nature in the same way that all stable physical matter necessarily consists of neutral atoms (more on that shortly). Likewise, the vertical axis is the magnitude and polarity of the waveform with anything above the real number line of one polarity (positive or negative), anything below of the opposite polarity and anything on the axis neutral. Again, positive and negative here don't mean good or bad, only opposite polarities in terms of the potential experiences that may be had and information that may be gained.

Next, the figure-8 shaped area contained between the upper and lower prime waves represents all possible potentials as a total probability space, while the specific geometry at each of the twelve base-12 positions determines the vibrational theme experienced at each discrete number position within that continuum. We will explore these numerology themes in depth in Chapter 5 and how we can use the base-12 numerology cycle as a personal road map of discovery and learning.

As discussed, position 6 represents the central point of awareness of the entire waveform, the origin of the *self*. Although position 6 occurs below the neutral reality axis, it still achieves neutrality as the intersection point between the upper and lower prime waves, essentially the point at which the polarity switches (picture a rubber band twisted in two). Also, because the base-12 cycle completes a full rotation from 0 to 12 (10 in base-12) with each cycle, it must reverse that cycle from 12 back to 0 to return to its original neutral state. The prime waveform it produces therefore follows a repeating figure-8 flow (indicated by the four black arrow heads along the boundary). This creates an inward flow from both directions above position 6 and an outward flow from below. This recirculating flow through position 6 is how each cycle ventures outward to gain new experience and then returns inward to integrate what was learned. Thus, with the completion of more and more cycles of experience, position 6 as the identity and consciousness of *love* expands its self-awareness – *love understanding itself better*.

In addition to the figure-8 flow path just described, the prime waveform also has a spin angle of momentum rotating into and out of the page (the four curved arrows). This, again, is because the prime waveform is not just a flat 2D figure-8 but actually a 3D double-helix spiral encoded into that 2D surface. And since the reality we perceive is a 3D holographic projection of that surface, the rotational spin of that twisting spiral is also perceived though not physically seen. How we perceive this spin depends on its direction: a forward spin as a sense of closeness and approaching toward us, a backward spin of separation and retreating away, an upward spin of allowance and lightness and a downward spin of influence and heaviness. This spin characteristic plus the height and polarity of each position within the figure-8 waveform is what gives each number its vibrational personality or *theme* in numerology – how each number intuitively feels and behaves.

Indeed, the base-12 prime waveform certainly does appear to be an excellent model of consciousness. However, it differs in one important way from our conceptual definition of consciousness. Instead of having a symmetric figure-8 shape perfectly aligned with the neutral reality axis, the prime waveform is *asymmetric* with more of a butterfly wing pattern – dipping below the neutral reality axis at its center and rising above at its two extremes. This places its point of awareness, position 6 of *love*, below the reality axis and therefore beyond the reach of physical perception.

But doesn't that make sense with how we actually perceive consciousness and love? We experience our sense of self as an entirely non-physical aspect within. The same goes for our sense of love as a feeling also emerging from within and just as subjective and illusive to grasp, sometimes frustratingly so. This only further supports the idea that love and consciousness not only emerge from the same point of awareness and in the same way but indeed *are* one and the same. Love is the origin, the journey and destination of what we perceive and how we perceive it. Love is literally All That Is in the eternal process of becoming.

Another important insight is that it's only the holographic projection of the prime waveform that presents to us as a lopsided figure-eight in our reality. The double-helix spiral operating in the background is perfectly symmetric and not relative to any physical frame of reference. It is therefore only in our limited physical perception of reality that the true nature of love and consciousness are concealed, just out of reach. Once we are no longer bound to physical form (i.e. between lives), however, we escape that illusion and regain full knowledge of the loving consciousness we are.

The 12D Crystalline Memory of Consciousness

The base-12 prime theory as presented thus far describes our physical reality plus time emerging from a 2D figure-8 projection of a 3D double-helix waveform. And all of this is based on the simple rotation of prime number positions 7 and 11 around the base-12 circle.

Recall, however, that we initially required all twelve number positions of the base-12 cycle in order to determine the four positions (1, 5, 7, 11) where the primes recur as a pattern. Further, it was through the vibrational combination of those four prime positions that reduced the pattern down to just positions 7 and 11 and created the two helical waves of the base-12 prime waveform. So although only two of the twelve positions define the twisting plane of probability that projects as 3D space and time in our reality, all twelve are necessary for that construct.

What this indicates is that the base-12 prime waveform itself emerges from a twelve faceted crystalline structure. In other words, the 7-to-11 probability plane rotates through twelve 30-degree reorientations, each representing a different plane of perspective and higher frequency of vibration than the one before. Still, it is only the 2D side view of this structure we can physically perceive. To help visualize this, the prime waveform and base-12 cycle are shown again below along with the twelve spin orientations of that cycle – only the side view of which we would perceive as if looking from the right (depicted by the little silhouette of a person).

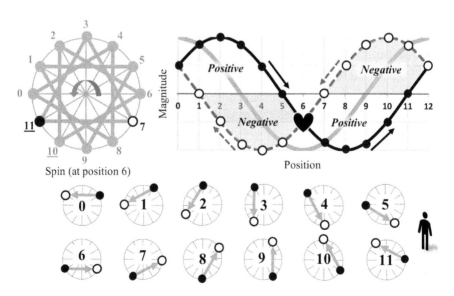

12D Crystalline Structure of Consciousness

As physical reality is patterned by the 2D side view of the prime waveform, we cannot directly perceive the spin angle of each of its twelve spin orientations but only their projected heights and relative positions within the cycle. This causes physical space to expand and contract according to the undulating geometry of the waveform. Also, despite the extra-dimensional spin angle not being directly projected into physical reality, its spin (or angular momentum) is nevertheless retained as an inherent holographic property of physical space and matter.

If, however, we could perceive all twelve spin planes at once it would be like looking along the length of the waveform in both directions from position 6. This would reveal the perfect crystalline structure shown to the left of the waveform. As this "end view" aligns the neutral reality axis with our line of sight, it removes the perception of sequence and time and effectively collapses the entire waveform back into its original timeless form as the base-12 circle. This therefore I suggest is the underlying cross sectional geometry of extradimensional space as a probability field of consciousness. This perhaps is why String theory is one of the more promising theories of physics today. It too builds upon the idea of the particles and forces of nature as emerging from vibrating strings and loops. Only, String theory views the different particles as unique individual strings rather than localized vibrations at various positions of the same figure-8 string.[17]

A perfect crystalline structure such as the one I am proposing also has a number of special properties, two of which relate to our discussion here. The first is that it is a theoretical requirement of a stable universe such as ours (technically speaking, because its ground state – position 6 in our case – would have zero entropy, or disorder, at absolute zero according to the third law of thermodynamics).[18] The second important property of crystals is that they are able to store information as memory, the same principle behind holograms. If everything is vibrationally structured through this crystalline structure including consciousness then it follows that everything should have this inherent crystalline memory and not just a fleeting awareness immediately lost.

Let's explore a few of the surprising ways that conscious memory is a very real property of physical phenomena.

The Crystalline Memory of DNA and Water

As discussed earlier, the DNA molecule reveals the double-helix structure of the prime pattern as closely as any physical form, with the two physical strands forming the chemical spirals we can see. If DNA is actually twelve stranded then there are an additional ten multidimensional strands we don't see. DNA having a crystalline structure does intuitively make sense as each DNA molecule is the energetic record of the complete human body and serves as the perpetual memory of our genetic code.

Snowflakes too mimic the base-12 multi-helical pattern beautifully though in a unique way. Rather than mimicking the full 3D double helix shape of the prime waveform, the typically six-pointed snowflake resembles just a cross-sectional slice. It's almost as if the prime waveform is literally frozen in time at the precise moment the water molecules freeze, capturing only a narrow slice of the prime waveform's full nature. However, the specific form a snowflake takes can be directly influenced by human intent and emotions.

The brilliant work of Dr. Masaru Emoto demonstrated how ice crystals when formed in a benevolent environment such as in nature or when subjected to loving words or peaceful classical music create elegant and symmetric snowflake patterns.[19] Conversely, when crystals are formed while subjected to hateful words, pollution or harsh heavy metal music (or anything sung by me), the snowflakes become deformed into distorted non-crystalline shapes.

Not surprisingly, the molecular structure of the water molecule itself mimics the asymmetric figure-8 shape and polarized neutrality of the prime waveform: of two hydrogen atoms bonded to one oxygen atom in a V-shaped configuration such that the upper part of the V (the hydrogen atoms) is slightly positive in charge while the lower part is slightly negative, resulting in a neutral H_2O molecule overall.[20] And, just as this upper-lower polarity of the water molecules enables them to attract each other easily despite their overall neutrality, so too does the polarized structure of the prime waveform enable it to attract like vibrations while maintaining its neutral sense of self.

As such, both DNA and water molecules are basically memory storage and retrieval devices along with all the atoms of which they are formed. With each of us being comprised of trillions of cells with each cell containing 46 DNA chromosomes and our body being over 60% water, we are powerful energetic memory banks indeed. Picture your DNA as a virtual antenna and recording device, retaining the memory of all your current and prior life experiences. This is often referred to as our *akashic record* in esoteric circles.

The Crystalline Memory of Significant Experiences

In the same way, the earth itself remembers the history of what has happened both globally and locally, taking on an energetic imprint of the experiences to which it was exposed. This is why those who are sensitive to energy can often feel the difference between walking through an ancient sacred space of worship and a former battlefield. The land vibrationally speaks of that which it has experienced.

This energetic imprinting I believe is responsible for much of what is reported as ghosts and hauntings. A significant event of some sort occurred in that location which was traumatic or otherwise dramatic enough to leave a limited recording of itself behind. Although such imprints do have a residual consciousness that seems to behave like an interactive entity, they are more like an audio or video recording stuck on replay. This I suspect is why many such hauntings are very repetitive in nature; the same type of sound, voice or visual apparition recurring again and again.

On a more personal level, traumatic events we experience individually are also prone to being frozen in time. Powerful events such as an accident or illness, emotional or physical abuse or recurring nightmare can become stuck in our consciousness and subconscious, playing again and again. And each time we replay it with the same emotional response, we essentially refreeze it into the present with the same dissonant vibe.

However, just as "traumatized" snow flakes can be melted and refrozen with a new imprint of loving energy, so too can past hurts and fears. Although we instinctively try to avoid our fears, this only serves to prolong them. This keeps our fears energized in our consciousness and attracting more of the same into our reality. Although we may feel we are over them we are merely supressing them as they continue to simmer away in the background with the same negative theme perpetuating itself.

I suggest the remedy is the opposite; to bring our fears into plain view, feel them honestly and examine them with objective eyes. Although this takes courage, it can be transformational. When we acknowledge past events as something we intentionally brought into our experience in order to expand our consciousness concerning a particular theme, we can better appreciate the underlying value of the experience and let it go. Perhaps the abusive relationship was to teach us not to give our power away, the illness to instill greater compassion for ourselves or the recurring nightmare to highlight a past-life karmic fear we have yet to release.

As I have found when dealing with the "ghosts" of my own past traumas, the hurts of yesterday are truly just illusions we drag into the present. Still, we revisit certain residual fears more frequently than others as they have more yet to teach us. Once we see them this way, as growth opportunities and valuable lessons, we can move on and reprogram the present with new beliefs such as replacing a sense of lack with abundance, unworthiness with self-worth, loneliness with empowerment or fear with courage. In this way we can break free from old patterns which no longer serve us and energetically advance. What we are actually doing, I suggest, is simply recalibrating and retuning our consciousness back in resonance with the frequency of love.

And that I believe is what is occurring at all levels of physical reality too, not just pure consciousness itself. Everything we perceive – from atoms to molecules, cells, living organisms, stars, galaxies and the entire universe – are conscious expressions of love seeking resonance. But if that's true, why does reality emerge in such distinct incremental steps with a whole lot of space in between? The answer to that riddle is perhaps revealed by the fractal nature of the base-12 cycle.

The Fractal Emergence of Love

One of the most fundamental and beautiful properties of nature is the way patterns tend to repeat at many levels of scale. These self-similar patterns or *fractals* can be found in numerous natural phenomena from clouds to cauliflower, ferns to feathers, lichen to lightning, snowflakes to sea shells and river tributaries to trees.[21] However, fractals are not just limited to physical objects but also occur with light, sound and anything vibrational. This therefore involves musical principles such as frequency, octave, harmony and resonance.

An octave of a vibration has twice its frequency and a harmonic any multiple higher. So each higher octave can be considered a self-similar fractal or harmonic of the fundamental vibration. A high C in music, for example, sounds similar to a low C despite them being octaves apart in pitch. Mathematically, an octave represents a 2:1 ratio of frequencies – the same 2:1 ratio of duality built right into the geometry of the base-12 prime cycle itself. The ancient Greek philosopher and mathematician Pythagoras, coincidentally the same luminary who gave us Pythagorean numerology, was the first to understand the science of music in terms of whole number ratios of frequency.[22] You rock, Pythagoras!

Being a simple vibration, the base-12 prime waveform follows the very same vibrational principles with each octave being a harmonic multiple of itself. And just like the chromatic scale of music, each vibrational cycle of the prime waveform represents an octave of twelve distinct frequencies or notes. As the prime waveform is produced by a full rotation of the base-12 cycle, a new fractal should therefore occur with each additional rotation of the cycle. However, as the base-12 prime waveform describes both the real (particle) and imaginary (wave) aspects of nature, the base-12 cycle would need to be expressed in terms of the bilingual real and imaginary language of complex numbers.[23]

Despite their name, complex numbers actually provide a simpler and more efficient way of describing nature than real numbers can alone. Also, there are a few different types of complex numbers depending on how many dimensions you are dealing with. Because our base-12 cycle and its double-helix waveform exist in one-higher dimensional space than our familiar 3D physical universe, we use the 4D version of complex numbers called *quaternions* (not to be confused with a square dance move that makes you dizzy). Quaternions are perfect for describing 3D rotations in 4D space which is exactly what the prime waveform is.[24] As they handle 3D rotations so naturally, quaternions find many practical applications not only in math and physics but also other areas such as computer animation and aircraft guidance control. So every time your plane lands safely, you can thank quaternions – even if you didn't realize you already landed because you were covertly playing video games on your phone, also thanks to quaternions.

Another property of quaternions is that rotations involving equal increments, such as the twelve number positions of our base-12 cycle, follow a logarithmic power scale. This means that a full rotation or octave of the base-12 cycle would consist of twelve rotational increments from 10^0 at position 0 through to 10^{11} at position $\underline{11}$. This in turn means that each successive octave or fractal of the base-12 prime pattern should hypothetically occur at 10^{11} multiples of each other. In other words, each larger fractal of the prime waveform pattern would automatically emerge every 10^{11}, or 100 billion, of the next smaller fractal down. Remarkably, this seems to agree with what we see in nature.

There are an estimated 10^{11} galaxies in the observable universe, 10^{11} stars per average galaxy, 10^{11} neurons in the human brain and 10^{11} atoms per DNA molecule, to name but a few. Keep in mind that those fractals small enough such that we are *outside* observers would reveal the full toroidal shape of the base-12 prime waveform. For the larger fractals such as our own galaxy and the entire universe we instead are *inside* observers. From that limited insider's perspective, we only perceive our galaxy as a rough Milky Way band of stars and the entire universe as a seemingly random expanse of galaxies in all directions.

As when watching TV, zoom in too close and all you will see are a meaningless bunch of pixels of different colors on a flat screen, zoom out too far and the entire TV itself will look like a little box of light, but sit at just the right distance away and your favorite show comes into view – including all those captivating pizza and beer commercials which ultimately give life meaning. And now that you can make out the relative movement of the 2D images on the TV, your mind perceives them as 3D and real. But sadly, the pizza and beer remain mere illusions.

Now that we understand why the base-12 prime pattern repeats at natural fractals of itself, we can appreciate how everything emerges as harmonized octaves of love. Let's now visit each of these octaves from largest to smallest, beginning with the biggest of the big: the universal.

The Emergence of Love at The Universal Scale

Here at the universal scale, we are going to see the base-12 prime waveform in action and just how well it describes the cosmos. And we're in for a bunch of surprises too, so buckle up. We will discover that the universe we call home is perhaps just one half of a cyclical universe/anti-universe pair and that the portion we physically perceive is just half that again; the upper left quadrant of the figure-8 where matter is possible. This includes a picture of the Big Bang as one of infinitely many polarity reversals between successive universe cycles of a toroidal multiverse. It explains the co-creation of matter and anti-matter and the relative excess of matter responsible for the physical make up of the universe. It also suggests that the "dark energy" of the seemingly accelerating expansion of space is just an illusion caused by the extra-dimensional spin of the base-12 prime waveform. This further suggests where we are in the current universe cycle today and why time is an illusion.

To help us navigate this far-out picture of the cosmos, we will walk through a typical life cycle of such a universe. But first we need to understand why we can only see a quarter of that picture and how this gives both the illusion of forward moving time and creates an excess of matter over anti-matter so that physical matter such as doughnuts and us can even exist.

Our One-Quarter View of Reality

In our earlier picture of the figure-8 cycle of consciousness on page 36 we saw how the neutral reality axis and the polarity reversal at position 6 divides that cycle into four quadrants. The upper left and lower right quadrants are regions of *positive* potential and the lower left and upper right quadrants of *negative* potential. Again, these four quadrants represent how love experiences itself through its full range of polarized potentials while still achieving neutrality overall. In the mathematical language of the prime vibration, these polarized potentials represent probabilities and are the foundation of modern physics.

However, in the tangible of world of physical reality and matter, only positive probabilities are possible. This means that all matter including us is limited to the upper left and lower right quadrants while the other two quadrants must be regions of anti-matter in order to achieve neutrality. Also, as we reside within this universal fractal of scale, our perception is further limited to just one side of the polarity reversal of position 6. This I suggest is "where" we are and the physical universe we experience.

As the upper left and lower right quadrants are flipped versions of each other but otherwise identical, we can pick either one as being the region of matter in which we reside. For illustration purposes, I have selected the upper left quadrant as our home address as shown below. This will serve as our blueprint for the life cycle of the cosmos.

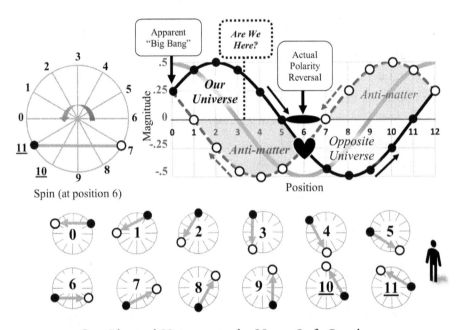

Our Physical Universe as the Upper Left Quadrant

Delayed Perception of Cycle Beginning and the Illusion of Time

This upper-left quadrant view of physical reality has a number of further implications. The first concerns how we perceive the manifestation cycle. Although each cycle actually originates from the polarity reversal at position 6, we perceive it as physically beginning at position 0. This is because the initial backward flow of the probability boundary from position 6 to position 0 occurs hidden below the reality axis, only emerging into physicality when it passes up through the reality axis at position 1 to create what appears to be a singularity at 0.

Next, as position 0 is where the flow direction of the probability boundary switches from backward (from 6 to 0 along the lower boundary of anti-matter) to forward (from 0 to 6 along the upper boundary of matter), we only perceive the forward progression of the upper boundary from 0 to 6. This gives the illusion of linear time as moving forward only. Nevertheless, everything exists in the present at the fixed central point of awareness of position 6, with the past and future simply illusions of our limited perspective.

Also, as our perception of positive probability as physical reality is interrupted by the hidden portion of each cycle which passes through negative probability, this "strobe effect" would create a sense of interval we perceive as continuous distance and time. In the same way a movie appears continuous, it is actually a series of individual frames projected rapidly enough to appear continuous relative to the frequency range we can perceive.

Initial Imbalance of Matter Over Anti-matter

We also see from the above geometry of our universe that matter and anti-matter are created in equal balance because of the overall symmetry of the base-12 prime cycle. That is, the two unshaded regions of matter are equal in area to the two shaded regions of anti-matter. However, each half-cycle taken on its own has an asymmetric geometry akin to a single butterfly wing.

This asymmetry affects the relative balance of matter versus anti-matter as we progress from position 0 to 6. The imbalance begins right away from position 0 to 1 with the probability window entirely above the reality axis with therefore only matter possible. This creates an initial excess of matter. From positions 1 to 3, with the probability window now spanning above and below the reality axis, the ratio of matter to anti-matter evens out to an equal 1:1 proportion. As matter and anti-matter annihilate each other 1:1, position 3 is the stage at which no further excess matter is created. Then from position 3 to 5 the ratio increasingly favours anti-matter until, from 5 to 6, only anti-matter remains. Thus the initial imbalance of matter is now reversed in preparation for the start of the next cycle through the polarity reversal at 6; the ending anti-matter of one cycle seeding the initial matter of the next through the 6 of love.

The Life Cycle of a Cyclical Universe

The current energy density make up of the universe is estimated at approximately 68% "dark energy", 27% "dark matter" and 5% regular matter.[25] We had best define these first so that we can see what the base-12 prime theory has to say about all of this, especially as we're going to end up taking pot shots at the two biggies.

Dark Energy: The hypothesized invisible outward *pulling* energy that causes the universe to appear to expand, as inferred by the increasing red-shift of distant galaxies.

Dark Matter: The hypothesized invisible matter *pushing* on galaxies to exhibit constant rotation like solid discs, as inferred by the gravitational bending of light around such structures.

Regular Matter: All visible matter, consisting of atoms from the periodic table of elements, such as galaxies, stars, planets and us.

The upper left quadrant of the base-12 prime waveform represents the six stages of our physical universe's life cycle from position 0 through position 6. These positions are not linear increments in terms of time but rather distinct developmental epochs through which the physical universe evolves. The physical size of the universe at each position is determined by the vertical magnitude between the upper and lower probability boundaries, although we only perceive the portion above the reality axis with the anti-matter portion hidden below. Also, the angular momentum or spin of the 7-to-11 probability plane "into the page" determines the inherent spin of space, what we incorrectly perceive as a linearly accelerating and decelerating expansion of space. So let's start our cosmic journey with the prime waveform as our guide.

Epoch 6 – Universe at Ground State, Polarities Reverse, Actual Start of Universe Cycle

Although the physical universe appears to emerge at position 0, we need begin our discussion at position 6 with the polarity reversal that actually marks the birth of a new universe cycle. Here, the 7-to-11 probability plane projects as a point singularity although it is really the intersection where the two prime wave boundaries overlap. As an intersection of two oppositely polarized probability boundaries, position 6 is neutral in charge with zero spin and would represent the ground state of the universe, what physics calls the Higgs field.[26] Occurring below the reality axis, this polarity reversal event is hidden from view. The cycle then follows its figure-8 flow along the lower probability boundary from position 6 to position 0, still hidden below the reality axis until it emerges into positive probability space from position 1 to 0.

Epoch 0 – Apparent "Big Bang" Beginning of Universe and Time, Temperature at Maximum, Universe Expanding Most Quickly

At position 0, the flow direction of the probability boundary switches from backward (right to left) to forward (left to right) marking the point at which we perceive the universe as beginning. Again, as the balance of the cycle from position 0 to 6 follows this forward trajectory we experience it as linear time moving forward only. However, this is an illusion as the entire figure-8 flow of the base-12 prime waveform is relative to its central origin and timeless state of love at position 6.

Although position 0 is a point probability like position 6 (i.e. edge view of the 7-to-11 probability plane), it is not a polarity reversal where the upper and lower boundaries overlap into a singularity as they do at

6. However, it would appear that way in physical terms. 3D space would be subject to the intense conditions of a singularity-type compression event, a "Big Bang" for all intents and purposes. This would create extreme pressure and heat which is confined within the point sized profile of the 7-to-11 probability plane. This supports the idea that all known forces are unified at position 0, if they were indeed separate forces. However, as the base-12 prime waveform is a single waveform describing everything, this would suggest the electromagnetic force is the *only* force and position 0 is the point from which that single force first manifests physically.

Also, recall that the thick solid grey wave that passes through the figure-8 profile of the base-12 prime waveform is the single cosine wave formed by adding together the two prime waves. This combined wave therefore indicates the net polarity or *instability* of each position in terms of probability. At position 0 we see that the grey wave is at its maximum amplitude of 0.5 or 50% indicating that the universe is at its most polarized and unstable, though positively polarized in favour of creating matter. Note too that position 6 is equally unstable but neutral in polarity, enabling the creation of both matter and anti-matter and thus perpetually powering the base-12 prime waveform.

Epoch 1 – Electromagnetic Force and Gravity Emerge, First Matter Forms (but no Anti-matter), Universe Expands Less Quickly, No "Dark Matter" or "Dark Energy"

As we move from position 0 to 1, the early universe does not yet touch the neutral reality axis. This is therefore a phase in which neutral matter is impossible but at which the electromagnetic force can take effect with the creation of massless photons as its force carrier particle and the electron as the simplest charged free particle subject to that force. So, this is where mass can first be imparted to particles by the Higgs field. Assuming the base-12 prime waveform is strictly electromagnetic, the photon's path is the entire figure-8 probability boundary of that waveform itself.

Position 1 is the first point in the universe cycle at which the probability window between the two prime waves touches the reality axis where constituents of neutral matter can form, just as 1 is the first prime position of the base-12 number cycle at which the primes are generated. As the probability geometry at position 1 is characterized by a singular polarity above the neutral reality axis, this is where the simplest charged constituents of stable matter can form, the up and down quarks.

Position 1 is also the point at which the angular momentum or spin of the 7-to-11 probability plane begins to angle downward towards the neutral reality axis as the plane rotates back. According to general relativity, the local force of gravity is equivalent to that felt in an accelerated frame of reference.[27] Also, as rotation at constant speed is equivalent to centripetal acceleration, this downward spin may actually be the cause of gravity. Because of the varying tilt angles of the 7-to-11 probability plane, the magnitude of the downward spin vector grows from zero at position 0 (plane horizontal) to fully downward at position 3 (plane vertical) and then decreases back to zero at position 6 (plane horizontal again). This would create a tapered gravitational effect which is strongest at position 3 due to its probability plane being vertical and fully projected (i.e. highest Higgs excitation). This in turn would imply there's no missing particle of gravity or "dark matter".

The rotation of the 7-to-11 probability plane would also drive the size of the universe as it is the tilt angle of the probability plane in 3D that determines the projected height of the probability window in 2D. So, like gravity, the universe expands from its smallest at position 0 to its greatest at position 3 and then contracts from position 3 to position 6. Also the rate of that expansion or contraction would vary from one position to the next in the same manner as the height of a sine wave generated by a rotating circle: expanding quickest in its first 30-degree increment (i.e. rapid inflation stage), less quickly in its second and least quickly in its third and then reversing that process by contracting least quickly in its fourth, more quickly in its fifth and most quickly in its sixth. Here at position 1, the universe is thus expanding less quickly than the rapid cosmic inflation of position 0.

Note too that position 3 marks the point at which expansion stops and contraction begins, where the universe is at its largest and most stable. Thus, there is no "dark energy", only the undulating probability window of the base-12 prime waveform. What we currently perceive as the accelerating expansion of space as measured by the "red shift" effect of light moving away from us is likely just that same effect caused by a rotating universe. And as we are within that rotating frame of reference, we wouldn't be able to tell the difference. That would make the red shift we currently assume as proof of accelerating expansion a red herring of cosmic proportions! So if "dark energy" and "dark matter" don't exist, the universe only contains real matter and would put to rest two of the biggest mysteries of physics.

As discussed earlier, the geometry of the probability window at position 1 would also explain why an excess of matter currently exists over anti-matter.[28] Only charged matter is possible at position 1 as the probability window is fully above the reality axis. This creates an initial excess of matter which is only rebalanced by the end of the universe cycle at position 6. This is apparent in the relative areas above the reality axis (matter) and below (anti-matter) only becoming equal over the entire cycle from position 0 to 6 due to the left-to-right asymmetry of those two areas.

The initial excess of matter and rotating geometry of the prime waveform would also explain why black holes occur in our universe. Recall from our earlier discussion on page 30 that a negatively curved universe such as one described by the prime waveform would be considered an AdS (anti-de Sitter) space. Again, this is a finite universe with an outer boundary which curves in upon itself. Another property of this type of universe is that the introduction of any additional matter would gravitationally destabilize it and result in black holes as a means of restoring stability.[29] As each half-cycle of the prime waveform is inherently imbalanced with an initial excess of matter (or anti-matter), the prime waveform predicts black holes (and anti-black holes) as an inevitable feature. The prime waveform is therefore a perpetual motion machine which converts matter into black holes and back again as seed matter for the next universe, conserving all energy and information in the process.

Epoch 2 – Matter Polarizes, Gravity Increases, Protons and Neutrons Form, Universe Expands Less Quickly

Position 2 marks the point at which polarity begins, with the probability window now spanning above and below the neutral reality axis. This is therefore the stage at which both matter and anti-matter are created and at which particle charges become polarized positive and negative. As the geometry of the probability window at position 2 now reflects a +2:-1 polarity ratio relative to the neutral reality axis, this would establish the +2/3:-1/3 charge relationship of the magnetic plane of the quarks. These include +2/3 charged up, charm and top quarks and their three -1/3 charged partners the down, strange and bottom quarks. As such, both the electrical and magnetic aspects of the electromagnetic force are now fully active at position 2 enabling the creation of all twelve subatomic particles (and their twelve anti-particles).

Now, the potential energy of the Higgs field is transformed into the mass of specific particles according to the projected exposure of the 7-to-11 probability plane which is in turn determined by the severity of its tilt angle. The universe at its hottest would therefore correspond to the greatest tilt angle from the Higgs field with lesser tilt angles as the universe cools. As such, the first quark pair to form would be the top and bottom quarks as they are of highest energy (mass) but also most unstable as a result, followed by the middle weight charm and strange quarks and then the up and down quarks being lightest and most stable. Likewise, the lepton pairs would form in the same manner; first the tau/tau neutrino, then the muon/muon neutrino and lastly the electron/electron neutrino. As such, by the point at which position 2 has reached its coolest, the lightest and most stable particles would remain: the up and down quarks and the electron and electron neutrino as the building blocks of all physical matter.

The +2:-1 probability geometry at position 2 would also require the quarks created there to combine according to this ratio. This would explain why the proton consists of two positively charged up quarks and one negatively charged down quark, while the neutron contains two down quarks and one up quark. This sets the groundwork for atomic nuclei to form.

Epoch 3 – Neutral Atoms Form, Gravity at Maximum, Expansion Stops and Contraction Begins, Temperature at Minimum

The probability geometry at position 3 now takes on a balanced ratio of +1:-1. This therefore marks the point at which the universe is cool enough for the +1 charged proton, neutral neutron and -1 charged electron to combine into neutral atoms. This enables the creation of the first and simplest element, hydrogen, from a +1 charged nucleus containing one proton surrounded by a single electron of -1 charge. Thus hydrogen becomes the basic building block of the periodic table of elements from which all physical matter is made. As neutral hydrogen atoms would not scatter photons of light as would the charged protons and electrons of epoch 2, epoch 3 would be when light from the early universe would first be visible as the electromagnetic radiation of what is called the Cosmic Microwave Background (CMB).[30]

As mentioned, position 3 is also the stage in the universe life cycle at which the universe stops expanding and is energetically neutral and stable. Gravity is also at its strongest here with the 7-to-11 probability plane vertical and fully projected. Since all stable matter assembles together at position 3 – from atoms to molecules, planets, stars and galaxies – this is where we feel the familiar fully downward gravity we have come to know and love (unless, of course, we happen to be standing on a bathroom scale). As this is the type of gravity we feel on beloved planet Earth, this is a big hint as to where we likely are today in our current universe cycle.

Note that the fully projected probability geometry of position 3 also results in the largest probability range of any position from 0 to 6. This would allow the broadest range of creative potential and manifestation diversity in the entire universe cycle. This, and the other above mentioned properties, makes epoch 3 the *only* stage at which full materialization may occur. Position 3 is truly the incubator of life and, numerologically speaking, the *catalyst* of complexity.

Epochs 4, 5, 6 – Reversal of Epochs 2, 1, 0

From the stability, neutrality, gravity and materialization of epoch 3, the universe shifts into reverse and starts contracting at an accelerating rate towards epoch 4. This causes the universe to heat up again and the atoms of all regular matter to separate back into free protons, neutrons and electrons. Matter further disintegrates from epoch 4 to 5 into its subatomic particles. However, a critical stage is reached at epoch 5; the reversal of polarity from that which occurred at epoch 1. This means that all remaining matter, now in its most fundamental constituents, has been fully annihilated by anti-matter. Thus, only anti-matter remains from epoch 5 to 6. Finally, the anti-matter reverses charge as it passes through the polarity reversal at epoch 6 to become the seed matter for the next universe cycle. The matter exiting from the polarity reversal would therefore be as evenly distributed across space as the anti-matter which entered, consistent with observations of the early universe.

So Where Are We Today?

To answer this question we need to establish a known reference point in the cosmological timeline at which a distinct measurable change occurred. This reference point occurred approximately 5 or 6 billion years ago when the previously decelerating expansion of the universe apparently began speeding up.[31] This finding was based on red-shift analysis of observations made of very distant supernovae by the Hubble Space Telescope in 1998, where the light from those luminous objects shifted towards the longer red wavelength of visible light as would happen if they were moving further away.

As position 3 is the stage at which the expansion of the probability window decelerates before contracting at an accelerating rate, this is presumably that same point in our cycle. It's reasonable to assume then we are just beyond position 3, as indicated with the little "Are We Here?" flag on page 48. This would agree with current cosmic conditions of a cool and stable universe abundantly materialized with neutral matter. A 2014 analysis of Cosmic Microwave Background (CMB) data from the Planck satellite confirms that the formation of large-scale galaxies is indeed slowing down.[32] This further supports our being just past 3 and that the galactic housing boom is over. Interestingly, our solar system is estimated to be 4.5 billion years old and was therefore formed near the very height of that housing boom.

The Gravity of Our Situation

If indeed we are just beyond the half-way point of epoch 3 in our current universe cycle, we will eventually begin experiencing somewhat more, shall we say, "tropical" conditions on our way to epoch 4. The good news, if you want to call it that, is that we and all physical life won't be around to endure the extreme temperature and pressure as all atomic matter will by that point have long since disintegrated back into its subatomic constituents. Yay for us!

However, only physical matter will break down, not the energetic and informational experience of that physical matter – our consciousness. So again I say, Yay for us! – but this time without the snarky sarcasm of one doomed to being vaporized in an insignificant puff of smoke. You know, like what happens to those poor expendable crew members wearing red shirts in *"Star Trek"* movies.

A Toroidal Multiverse Emerging from a Black Hole

When we consider the overall figure-8 geometry of the dual universe cycle, we see that following each four-epoch period during which matter can form (from 1 to 5 and 7 to 11) is a two-epoch gap when it cannot (from 5 to 7 and 11 to 1). Thus, the fundamental 2:1 organizing principle of duality is revealed once again and at a spine-tingling cosmic scale. This also offsets position 6 below the neutral reality axis and creates a two-position gap directly on the reality axis, flanked by a four-position wide probability region on either side.

As mentioned earlier, the twisted figure-8 path of the prime waveform follows a toroidal flow through the polarity reversal of position 6. As this flow originates and returns to that same central intersection point, every previous and subsequent universe cycle would share that same origin. Assuming that each new universe cycle is a unique expression from the one before and that there are infinitely many such cycles, the probability boundary of this multiverse would form the complete 3D surface of a torus (specifically, a "horn torus" which is a doughnut without a hole – the type I prefer). And in the same way the 2D figure-8 shape of a single universe cycle would become a 3D torus over infinitely many cycles, the 1D gap along the neutral reality axis between positions 5 and 7 would create a 2D circular void of probability that would appear in physical space as a 3D spherical event horizon from which even light cannot escape. This I suggest is the nature of all black holes including here at the universal scale as well as at the galactic and atomic scales.

Nevertheless, our frame of reference being from within one half of the dual universe cycle would make it impossible for us to perceive the full toroidal shape of the universe or its central black hole. This is because it is in the future relative to our linear perception of light and time. Nevertheless, this model suggests that each universal half-cycle is an opposite expression of the other half, interconnected or entangled with each other and every other prior universe through the shared origin of the 6 of love.

Shedding New Light on the Speed of Light

The geometry of the base-12 prime waveform also suggests new insights into the nature of light. As the waveform is conjectured to be strictly electromagnetic under this theory, the photon as the carrier of the electromagnetic interaction is what defines the speed or kinetic energy of the figure-8 probability boundary. Because the photon is massless, its kinetic energy or energy of motion equals its total energy.

The fundamental frequency of the base-12 prime waveform is 1 as its single wavelength spans one cycle. Also, its wavelength is 2×12^6 because the quaternion power cycle for the horizontal position axis spans from 12^0 to 12^6 and back again. As the speed of light c is equal to frequency f times wavelength λ, we obtain $c = 1 \times (2x12^6) = 2 \times 12^6$. As such, the speed of light in terms of the dimensionless logarithmic position scale of the prime waveform is 2×12^6. However, as our physical perspective is limited to only a half-cycle of the waveform (the upper left quadrant from position 0 to 6, representing matter), what we perceive as the speed of light is 12^6.

Now, to relate this dimensionless theoretical speed of light of 12^6 to the conventional units of *metres per second*, we need to characterize both under a common baseline of units of distance per unit of duration. In the context of time, a full cycle of seconds is a minute, a cycle of minutes is an hour and a cycle of hours is a clock cycle. As a completed cycle in positional notation equals 10 (i.e. 1 cycle of anything + 0 remainder = 10), a minute can be expressed as 10 second-cycles and an hour as 10 minute-cycles. Therefore, within one hour-cycle there are 10 minute-cycles x 10 second-cycles or a total of 10^2 second-cycles under this generic notation.

However, the logarithmic position scale of the base-12 prime waveform already represents hour-cycles in that it is directly derived from the base-12 clock cycle itself. This means that 12^6 positional "hours" x 10^2 second-cycles (which equals 298,598,400) is theoretically equivalent to the speed of light in metres per second, provided these units can be considered dimensionless as well. This is a valid assumption as the speed of light is a universal constant regardless of an observer's frame of reference (i.e. scale invariant). With the measured speed of light being 299,792,458 metres per second we find the theoretical prediction is remarkably close, within 0.4% of the observed value.

If valid, this conjecture raises the question as to why the observed speed of light is slightly higher than the predicted speed. I suspect this is due to the same reason that the universe currently appears to be expanding at an accelerating rate; the extra-dimensional spin is slightly amplifying the apparent speed of light towards us now that we are past position 3. As the 7-to-<u>11</u> probability plane at position 3 is fully perpendicular to our frame of reference, the trajectory of light along its probability boundary would not have any forward or backward angular momentum to distort its apparent speed. As such, position 3 would be the only point in the universe half-cycle at which the speed of light would be perceived accurately under this reasoning.

This alternative view of the speed of light also offers an intriguing interpretation of Einstein's equation $E = mc^2$ for the equivalence of energy (E) and mass (m). If we substitute 12^6 for the speed of light (c), we obtain the expression $E = m \times 12^{12}$. From the same property we used to derive the quaternion powers cycles, raising a quaternion to a power of 2 results in a rotation twice that of the original rotation. As 12^6 is equivalent to a 180-degree rotation half way around the position power cycle, raising 12^6 by a power of 2 to 12^{12} is equivalent to a full rotation of 360 degrees. As such, the kinetic energy of a particle of matter (or anti-matter) is equivalent to it moving through all twelve positions of the base-12 prime waveform. As a complete lap of the waveform's figure-8 probability boundary completes twelve positions forward and back, matter and anti-matter are necessarily created equally to accomplish this.

In a more general sense, $E = m \times 12^{12}$ also means that $m = E/12^{12}$ or that the mass of a particle at rest is equal to the potential energy of the base-12 prime waveform being grounded to a specific position along the neutral reality axis. This returns us to the idea that wherever the figure-8 probability boundary intersects the horizontal reality axis are where particles of matter manifest. As such, the prime waveform never actually "collapses" but rather, as David Bohm suggested, passes through our physical space-time plane of perception at distinct quantum locations.

This also suggests that the underlying reason why the speed of light is constant is because it is a fixed property (half cycle) of the scale invariant base-12 prime cycle. This further implies that position 6 as the origin of the waveform is the static zero-point potential of the Higgs field, while the waveform it generates is the dynamic expression of that potential. As the waveform is posited to be electromagnetic, this dynamic expression would therefore take the form of the massless photon travelling at the speed of light. In effect, position 6 is the 12^6 energetic potential of the Higgs field fully constrained, position 0 is that potential fully liberated as the kinetic energy of a massless photon and every position in between is partially constrained and therefore partially massive.

This presents light as something even more fundamental than just the physical speed limit of electromagnetism. It is the 12^6 half-cycle of the base-12 prime waveform that we can perceive, the half-cycle of matter. Light also traverses through the other half-cycle of anti-matter, of negative probability, but does so invisibly with only its strobe effect hinting at its true dual nature. As 12^6 is the universal energetic expression of the 6 frequency of love, we gain profound validation of the ancient spiritual belief that love *is* light.

It's appropriate we end our discussion on the cosmological scale with the topic of light. Without the energy and warmth of light to initiate and sustain biological life, the cosmos would be a much less interesting place – and less wise. Granted, consciousness would still exist, I believe, but would not have the profound opportunities for learning and expansion which only physical experience can provide. It is the diversity of challenges, spectrum of emotions and tension between personal identity and mortality which makes our souls' basic training on "Boot Camp Earth" so valuable. But only if someone turns on the lights first, which happens at the galactic scale of reality and the stars which make each galaxy possible.

The Emergence of Love at The Galactic Scale

From the cosmically humungous we now zoom in to the galactically still pretty darn big. Galaxies are the next smaller fractal of the base-12 prime waveform within that of the entire multiverse, a fractal the base-12 prime theory suggests inherently occurs at increments of 10^{11} in terms of particle density. So where the observable universe takes shape from its 10^{11} galaxies, each average galaxy owes its structure to its 10^{11} stars.

"Dark Matter" as the Tapered Gravity of Extra-Dimensional Spin

A longstanding mystery in cosmology is why galaxies rotate as they do. Rotating much like fixed discs, all their galactic matter of stars and planets rotate about their centers at a constant rate rather than the outer bodies spinning much more slowly as would be expected. This odd behaviour was identified in 1933 when astronomer Fritz Zwicky noticed the visible mass of galaxy clusters could not account for their motion.[33] So he coined the term "dark matter" as a label for what he thought must be some type of invisible matter – and a lot of it - responsible for holding galaxies together this way. Then, in the 1960's, the astronomer Vera Rubin further confirmed the constant nature of galaxy rotation through an extensive study of many individual galaxies. This is the typical though baffling behaviour of all spiral galaxies.

If the base-12 waveform were responsible for the equalization of velocity across the profile of a galaxy, then it would need to induce faster speeds towards the perimeter of the galaxy without the outermost matter jettisoning off into space due to centrifugal force. This is exactly what the base-12 prime profile reflects. The incremental tilting of the 7-to-11 probability plane would cause the angular momentum of extra-dimensional spin "into the page" to be at its maximum at position 0, decrease to zero at position 3 and then increase again to maximum at position 6. This tapered gravitational effect would therefore increase rotational velocity towards the perimeter of a galaxy (position 0) and its central black hole (position 6) while in between (at position 3) the angular momentum would be directed fully downward towards the reality axis with only the attractive force of conventional gravitational being felt.

This intensification of spin at the outskirts of galaxies may also be responsible for the comfortable separation which exists between adjacent galaxies, helping to distribute matter evenly throughout the universe. This may in turn help explain recent findings that all large-scale matter in the universe, such as galaxies and galaxy clusters, tend to concentrate along what appear to be filaments of a giant interconnected cosmic web.

As the base-12 pattern acts as a continuously repeating waveform propagating outward in 3D physical space from each black hole, it would follow that all galaxies it produces would naturally seek to align side-by-side in a somewhat linear fashion. Each filament of galactic matter would eventually cross paths with another filament and want to reconnect according to the overall base-12 geometry, perhaps forming the beautiful lattice work observed.

In a 1998 paper by physicists E. Battaner and E. Florido, compelling evidence is cited that galactic clusters not only form such linear filaments but that adjacent filaments connect to one another in a specific geometric way at the very large scale of superclusters.[34] They seem to form an organized crystalline grid work of connected octahedrons, coined the "Egg-carton Universe" structure by the authors. This geometric patterning of galactic superclusters further supports the idea of an overriding geometric property to the universe such as the base-12 prime pattern.

The fractal nature of the galaxies occurring at a 10^{11} density of stars would also explain why this constant velocity disc-like behavior only seems to be a property of mature galaxies and galaxy clusters. A young galaxy still forming would not yet have reached the density threshold of

10^{11} stars required to fully manifest the base-12 prime pattern. Similarly, when a mature galaxy runs out of steam as its stars die and the black hole is weakened through radiation, we would expect to see that galaxy become too diffuse to take shape once more. And without the tapered gravitation influence of the prime pattern, little or no dark matter effect would be seen. This too agrees with actual observations of diffuse galaxies.[35]

At the galactic scale as at the cosmic, the prime waveform therefore suggests that dark matter and gravity are not separate forces or some invisible particles but rather the tapered toroidal influence of the extra-dimensional spin of the base-12 prime pattern. Galactic black holes are likewise smaller cousins of the universal black hole and follow the very same structure and behavior as we will explore next.

Black Holes as Gravitational Funnels

The toroidal framework of the base-12 prime model may also clarify the nature of black holes such as the one we know exists at the center of our own galaxy. As depicted on page 48 by the black ellipse over position 6, there is a region in the center of the torus between 5 and 7 that is beyond the spacetime probability boundary of the base-12 prime waveform but which nevertheless leaves a visible gap directly on the reality axis. This gap, which appears as a spherical black hole to us in our 3D frame of reference, I suspect is probabilistically a cone or funnel shape.

Why I suggest black holes are conical is because the toroidal geometry and flow of the prime waveform would require it. If the upper inflow which directs energy, matter and information into the center of the torus is the black hole we can perceive then there must be a corresponding outflow hiding from view to maintain energetic balance. We would need to be able to see multidimensionally, however, to observe this inner polarity reversal as it would be part of the underside of our spacetime plane where anti-matter resides. The visible black hole above position 6 and the hidden polarity reversal of position 6 below would thus work in tandem as a funnel-shaped dynamic of intense gravitational recirculation.

Now, building upon the idea of a toroidal flow along the probability boundaries of the prime waveform, it would be along these boundaries that mass is delivered throughout the waveform from the central mass concentration between positions 5 and 7. As such, this would imply that the total mass-energy of a black hole (or Higgs field, more on that shortly) is what determines the mass-energy available to be distributed as matter in a galaxy (or atom). Supporting this conjecture are the findings of a 2015 study of elliptical galaxies by a team at the Harvard-Smithsonian Center for Astrophysics confirming a direct link between the amount of "dark matter" found in galaxies and the size of their black holes.[36] This is in line with the premise that the black hole (or Higgs) may indeed be the repository of mass for the entire galaxy (or atom).

Another recent finding seems to support the idea of prime positions 5 and 7 being the conduit of concentrated mass into a black hole. In terms of the geometry of a black hole, positions 5 and 7 represent the *event horizon* threshold of the black hole. Again, this is the boundary where the base-12 prime waveform of physical reality dips out of sight into the anti-matter plane prior to each polarity reversal. As the conduit for mass between the matter and anti-matter planes and the probability boundary between those two states, we would expect the event horizon around any black hole to possess a high-energy ring of mass.

This is indeed what was measured in 2017 by the Laser Interferometer Gravitational-Wave Observatory (LIGO) following the merger of two binary black holes.[37] In addition to reconfirming the existence of gravitational waves as predicted by relativity theory, scientists also detected gravitational wave "echoes" which indicate the event horizon has an energetic structure. This highly anticipated structure was finally visually confirmed in 2019 by the Event Horizon Telescope when an image of a black hole was captured, with its bright orange accretion disc ablaze for all to see.[38]

Gravity as the Localized Effect of Extra-dimensional Spin

Gravity is observed as the pull between any massive objects that causes those objects to accelerate towards each other, a pull that seems infinite in range but decreases with distance. According to Newton's law of universal gravitation, the attractive force between any two massive bodies is proportional to the product of their masses and decreases inversely proportional to the square of the distance between them. This law was further refined by Einstein's theory of general relativity which also accounts for very strong gravitational fields such as in the vicinity of black holes and objects very close to each other. For all other situations, however, Newton's Law provides an excellent approximation to how classical gravity works.

Coupled with this localized gravitation between massive bodies is the other gravitational phenomenon we just visited, the varying gravitation due to the overall tapered profile of extra-dimensional spin. This, again, is the rotational intensification property which enables galaxies to rotate at constant velocity and which I offer as a possible debunking of "dark matter". So, together, these two gravitational influences may actually be what we experience as gravity.

Still, it is the local *downward* gravitation we mostly experience as the Earth and our own body mass are subject to the fully downward angular momentum of position 3. This is why Newton's classical gravity was the prevailing theory of gravity for so long, it is sufficient to describe what we personally experience in daily life. In fact, the inverse square law of classical gravity is built right into the geometry of position 3 as follows.

The 1:1 polarity ratio at position 3 is characterized by the upper and lower prime waves being equidistant from the neutral reality axis, essentially the projected side view of a circle. As this 2D circle in 3D space is actually a projection of the 3D surface of a sphere in 4D quaternion space, position 3 represents the geometry of a point source emanating spherically outward. And because the surface area of a sphere is $4\pi r^2$, the intensity of any such point-source radiation is inversely proportional to the square of the radial distance (r) from that source.

The gravitational pull between two bodies being proportional to the product of their masses also beautifully mirrors the nature of prime numbers in that every non-prime integer is the product of primes as factors. In this respect, individual massive bodies serve as the prime number "factors" of all gravitational interactions in the cosmos, with the product of their masses being their resulting mutual "integer" attraction.

So rather than gravity being a separate force unto itself I suggest it is simply a consequence of the extra-dimensional spin of the base-12 prime waveform, a core property within everything and at all scales. This is consistent with what Albert Einstein called the principle of equivalence in general relativity; that gravity and inertial acceleration have identical effects in terms of the force observed. Only if we know the state of motion of an object can we distinguish whether an observed force is due to gravity or acceleration. Consequently, I don't think there's any underlying particle involved in mediating the gravitational effect such as the much speculated yet thus far elusive "graviton".

Another important characteristic of gravity is that it is felt as a strictly attractive force, operating in one direction only, unlike other forces of nature such as electromagnetism which both attract and repel. But why is gravity a one way street? This I believe is answered once again by the geometry of the base-12 prime waveform.

Recall that our frame of reference of physical reality is limited to the upper left quadrant of the waveform, from positions 1 to 5 along the prime wave of matter. Referring back to the diagram on page 48, we see that the direction of spin (angular momentum) of the 7-to-11 probability plane for all these positions rotates downward towards the reality axis in varying degrees. This therefore makes our experience of gravity as being downward too. The same applies to the other side of the matter wave from positions 7 to 11 with spin directions this time all pointing upward towards the reality axis, although we don't simultaneously perceive that "anti-universe" reality.

Magnetism as the Toroidal Flow of the Base-12 Prime Waveform

Not only does a plausible explanation for gravity emerge from the base-12 prime theory as posited above but for magnetism too. However, where gravity is the singularly polarized effect of the extra-dimensional spin of one prime wave, magnetism is the double polarized interaction between both prime waves. As such, magnetism relates to the overall toroidal flow of the base-12 prime waveform.

When we look at the toroidal pattern of a full double-universe cycle we see that the toroidal rotation induces a specific figure-8 flow, an inward flow along the upper probability boundaries towards the inlet of the central black hole denoted by the upper two arrows on page 48 and an outward flow along the lower boundaries away from the outlet of the black hole below indicated by the lower two arrows. Presumably then, all matter in our universe is imbued with this inherent figure-8 flow dynamic.

Consider, for example, the toroidal magnetic field generated around any simple bar magnet. A magnet will exhibit this characteristic field when you lay it flat on a sheet of white paper and sprinkle iron filings around it. The lines of magnetic force will impart their pattern to the filings, extending out from one pole in a toroidal pattern and curving back towards the other pole. So even this most basic of magnetic fields conforms to the toroidal energy flow of the base-12 prime pattern, where the inbound pole of our magnet represents the point above the polarity reversal at position 6 and the outbound pole the point below. In this way, the magnet itself essentially acts as a crude physical representation of a black hole we can perceive and hold in our hand without getting sucked into oblivion.

The field around a bar magnet is just one of many real life examples of the toroidal flow of magnetism. Others include the magnetic field generated around any wire carrying electrical current, the planetary magnetic field around the Earth and the biofield around all living organisms. The key requirement for a magnetic field is the movement of charged particles. In the case of the base-12 prime waveform as an electromagnetic pattern those particles are photons of light and their path the figure-8 probability boundary. This brings us to the intriguing prospect of massless objects.

Manipulating Local Gravity through Phase Displacement of Magnetic Fields

If the overall toroidal flow (i.e. magnetism) of the base-12 prime waveform and the localized extra-dimensional spin (i.e. gravity) of positions along that waveform are inherent properties of the same waveform then changing one should hypothetically affect the other. That is, altering the vertical gap between the two prime waves should alter the size of the black hole above 6. And the most direct way to do that I suspect would be to change the phase displacement of the two prime waves relative to each other.

We discussed earlier how the very shape of the base-12 prime waveform is generated by a 120-degree phase shift between the two prime sine waves. It is this 120-degree phase displacement between the two magnetic prime waves which produces the specific magnitude and geometry of the black hole gravitational engine at the center of the waveform between positions 5 and 7 and the amplitude of the combined grey wave. If, however, we could alter the phase shift between the two prime waves, even just slightly, this would hypothetically alter the size of the black hole and therefore its gravitational effect. Increasing the phase displacement by 30 degrees for example, from 120 degrees to 150, would result in the modified waveform shown below.

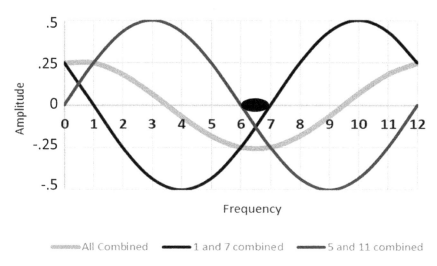

Increasing Phase Shift of Base-12 Prime
Wave from 120 to 150 degrees

We see that the width of the black hole shrinks in half, from two positions wide in our original unshifted waveform (from 5 to 7) to one position wide in our shifted version (from 6 to 7). However, the wavelength and amplitude of the overall waveform remains unchanged, only shifted to the right by 0.5 units. Note too how the amplitude of the grey cosine wave decreased by half (from 0.5 to 0.25). As this amplitude represents the magnetic potential or "strength" of the polarity reversal at 6, the effective strength of the gravitational effect is likewise cut in half.

This suggests that the gravitational effect on any matter subjected to such a modified pattern of magnetic fields should be reduced accordingly. That's because the mass density of the electromagnetic system is being effectively reduced by creating a smaller mass center without changing the space-time environment (wavelength and frequency) of the base-12 vibration. If the structure of all matter is indeed created according to the vibrational base-12 prime pattern, right down to the structure of the atom, this suggests that the atomic mass of matter can conceivably be altered too.

Note that increasing the phase displacement 30 degrees further would bring the two prime waves 180 degrees opposite one another. The two waves of equal amplitude would therefore cancel each other out and effectively eliminate the black hole and flat-line the grey combined wave. No black hole, no gravity. No gravity, matter becomes massless – in theory at least.

Although we can safely assume that the inherent prime pattern that creates the universe is unchanging and unchangeable, perhaps it can be artificially tweaked to a limited extent using magnetic fields in a controlled setting. Yes, I'm suggesting that the effective mass of objects should be able to be altered experimentally. With just the right arrangement of two overlapping magnetic fields that mimics the base-12 prime waveform geometry, gradually increasing the relative phase displacement of those magnetic fields should theoretically reduce the effective mass in the region local to the experiment. This presents exciting possibilities for massless objects and the incredible applications this would create if true.

An early pioneer of exploring the possibility of manipulating mass was Nikola Tesla. Although Tesla is best known for his development of alternating current electricity, he was a prolific and intuitive experimenter and inventor fascinated with electromagnetism and the quest for harnessing unlimited free energy for the world. One such invention was his famous Tesla coil for the wireless transmission of high voltage electricity.[39]

Featuring a pair of capacitors and spiral copper coils separated by a spark gap, the Tesla coil operates by inducing a magnetic field in the first coil which in turn creates an electrical current in the second coil as the voltage shoots across the gap. When the gap and timing are adjusted just right, this creates resonance between the two coils and a rapid oscillation of electricity back and forth. If it weren't for the electrical losses due to the heating of the air between the spark gap and the limited conductivity of copper, the Tesla coil could theoretically be a self-sustaining source of energy without need for an external power source to keep it running.

This suggests that confining the spark gap within a perfect vacuum and utilizing super-conductive materials could minimize the transfer loss between coils. And note how similar in configuration the double helix electromagnetic structure of the prime waveform is to that of the Tesla coil, including its two polarized "coils" separated by a central gap in the form of a black hole vacuum. If this device could be perfected perhaps it would not only realize self-sustaining electricity production envisioned by Nikola Tesla but also the localized gravitational control suggested possible by the prime waveform.

Nevertheless, Tesla was clearly hot on the trail of deciphering the base-12 prime waveform but simply ahead of his time and humanity's readiness for applying such technology effectively and responsibly. Based upon how strongly my intuition has been guided to uncover the base-12 prime waveform and to connect it with Tesla's earlier work, I suspect many others are also synchronistically being drawn to the prime waveform as we speak.

Such is the way it seems with technological advancement; once we are ready for new technology as a species that information becomes intuitively accessible to all through the collective consciousness. Case in point, various individuals around the globe were on the cusp of successful manned flight when the Wright Brothers achieved the first publicized flight. As such, I expect to see much more in the coming years regarding the base-12 prime pattern from many different fields of science as others connect the same dots through their own intuitive nudges.

The Emergence of Love at The Scale of Life

From the galactic we now explore the medium scale of biological life; of cells, molecules, plants, animals and us. This is the familiar level of nature we can experience directly with our physical senses and the help of a powerful microscope. We begin with arguably the most important component of life that defines the biological blueprint of all living things, the DNA molecule.

Base-12 Prime Structure of DNA and the Genetic Code

As touched on earlier, the three-dimensional double-helix spiral shape of the DNA molecule has a remarkable resemblance to the base-12 prime waveform. Consisting of 10^{11} atoms, this particular molecule meets the particle density requirement as a fractal of the base-12 prime waveform and therefore gets its good looks honestly. As we will discover in this section, this resemblance is more than merely skin deep.

Just as DNA contains the entire blueprint for making a living organism from scratch, the base-12 prime pattern appears to be the vibrational blueprint for the structure of DNA and the energetic formula of the genetic code itself. This is reflected in the way each amino acid is coded and how the four nitrogen bases combine to form a symmetric and consistently spaced double-helix structure. To explain, we return to our mathematical description of the base-12 prime pattern from Chapter 2.

To recap, the vibrational structure of the base-12 prime pattern stems from the fact that all primes (other than 2 and 3 as the factors of 12) occur at the 1, 5, 7 and 11 positions of the base-12 circle as a periodic cycle. Within each base-12 cycle of these four prime positions, every fourth prime candidate (PC_4) is related to the previous three by the equation $PC_4 = PC_3 + PC_2 - PC_1$. For example, $11 = 7 + 5 - 1$ just as $13 = 11 + 7 - 5$ or $17 = 13 + 11 - 7$. This relationship follows from the fact that the four prime positions vibrationally reduce to a pair of sine waves of equal amplitude 6, that is, $11 - 5 = 7 - 1$ or, as an equation, $PC_4 - PC_2 = PC_3 - PC_1$. We see that this is equivalent to our equation above, simply rearranged with the PC_2 term moved from the right side of the equation to the left. In the same way, the genetic code of DNA for creating amino acids involves the selection of three bases, called a codon or triplet, from four possible choices: adenine (A), thymine (T), guanine (G), cytosine (C) as depicted below.[40]

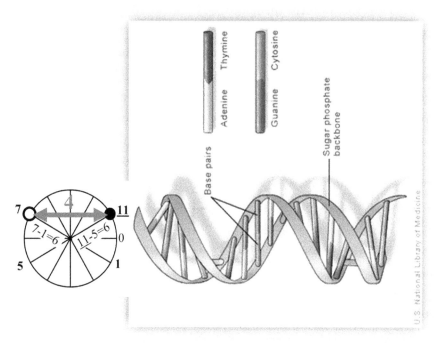

(Source: U.S. National Library of Medicine)

DNA Structure and Coding of Bases

Next, the way in which the base pairs combine in DNA to maintain a constant spacing between the two helical strands follows the constant spacing maintained between the two prime waves of the base-12 prime waveform. Recall that this constant spacing is the positional difference of 4 between positions 7 and 11 on the base-12 circle and which forms the fixed gap between the two helical prime waves. This is shown to the left of the illustration above as the grey arrow spanning positions 7 and 11.

Also, in the structure of DNA, an adenine on one chain is always paired with a thymine on the other and a guanine always paired with a cytosine. This pairing rule ensures that a smaller base always joins with a larger one so as to separate the two DNA strands evenly along their entire length. Further, this specific pairing is the only arrangement that works to bond the two chains together as it's the only configuration that fits together to form the hydrogen bonds needed to hold the dual strand structure together.

In the base-12 prime waveform we see this same property in the way the two net vectors of length 6 which define the 7-to-11 plane are each the result of pairing a smaller prime position to a larger one on the opposite side of the base-12 circle. The net vector of length 6 at position 7 is the result of combining opposite vectors of length 7 and 1 (i.e. 7 - 1 = 6), while the net vector at position 11 is from combining opposite vectors of 11 and 5 (i.e. 11 - 5 = 6).

The 120-degree offset between the two sine waves of the base-12 prime waveform also appears to be mirrored in what are called the major and minor grooves within DNA. The major groove is a functional feature of the molecule making it easier for DNA binding proteins to interact with the bases. Further, the two strands of DNA even follow the same anti-parallel flow as the two base-12 prime waves with the bases following opposite sequences on either strand.

In short, the DNA molecule is a functional and complete miniature version of the otherwise hidden 3D double-helix shape of the base-12 prime waveform, a compact blueprint that translates the mathematical plan of nature into physical form. Or in the loving words of Austin Power's nemesis Doctor Evil, "Mini-Me, you complete me."

Epigenetics and the DNA Phantom Effect

As we wrap up our discussion of the base-12 prime pattern revealed in DNA we should touch on the related topic of epigenetics. Epigenetics is a branch of biology that studies how outside influences control the behaviour of genes and DNA.

Dr. Bruce Lipton, a pioneer in the field and author of the groundbreaking book, "The Biology of Belief", showed that genes and DNA are not what primarily control our biology but rather the chemical and energetic environment outside the cell.[41] This includes the energy of our conscious thoughts, suggesting why positive thoughts can bring healing effects through the placebo effect, homeopathy and even spontaneous remission of disease while negative thoughts and stress can trigger or worsen illness. What Dr. Lipton essentially demonstrated was that DNA energetically communicates with its environment and that this environment acts as a conscious intelligent field.

There is in fact substantial experimental evidence beyond Dr. Lipton's work for the existence of a universal consciousness field and its direct connection to DNA. One particularly fascinating experiment was first performed by quantum biologist Dr. Peter Gariaev in 1985 and then jointly expanded upon in the 1990's by Dr. Gariaev and Dr. Vladimir Poponin. I first learned of this experiment through the book "The Divine Matrix" by Gregg Braden, another trailblazer merging science and consciousness.[42]

The phenomenon Gariaev observed, coined "The DNA Phantom Effect", was unexpectedly discovered when he was investigating the effect of DNA on photons, the particles of light. The experiment involved creating a vacuum inside a hermetically sealed tube and then introducing photons of light into the tube using a laser. His objective was to determine if the photons were scattered randomly within the tube, which precise measurements did indeed confirm.

However, when he then introduced a sample of human DNA into the tube and repeated the experiment, the photons immediately became ordered into a distinct pattern. Equally surprising was that when the DNA sample was removed the photons remained ordered in that pattern for days afterwards. So just the presence of a living DNA sample was somehow informing and organizing the quantum particles (photons) of the electromagnetic field in a lasting way.

This, Gariaev suggested, could help explain why subtle energy phenomena such as alternative healing can work. I personally believe this same effect is why psychic connections can be enhanced by holding an object with the energy of the subject, a practice called psychometry. I have experienced this myself during my mediumship readings. Holding an object of significance to the person with whom I am attempting to connect, such as a piece of their jewellery or photo, often strengthens my connection for clearer messages.

I suspect the DNA pattern resonates so strongly with photons of light because DNA is so close in form to the base-12 prime waveform spiral itself. Indeed, DNA is perhaps the closest physical manifestation of the prime waveform we can find in nature. This, by the way, suggests the DNA double-helix may also be multidimensional and possess the same 12D crystalline structure of the base-12 prime pattern; the two physical strands we can see and ten energetic ones we cannot.

We truly are energetically more powerful than we know and the double-helix base-12 prime pattern seems to be the organizational template for translating the energetic into the physical in the most efficient and balanced way, through our DNA.

Base-12 Prime Structure of Neurons and the Human Brain

Like the 10^{11} atom density of DNA, the human brain has a 10^{11} neuron density.[43] True to form, the structure and operation of the human brain appears to follow the base-12 prime pattern faithfully. Not only does the brain have two symmetric hemispheres which communicate to each other through a central connection (the corpus callosum), but each hemisphere also performs opposite cognitive functions; the left side specializing in the tangible, practical and logical (matter) and the right side specializing in imagination, creativity and intuition (anti-matter).

Further, located where the two hemispheres merge is the pineal gland which modulates daily and seasonal sleep cycles. So, just as the polarity reversal at position 6 controls the oscillation of the base-12 prime waveform between the physical and non-physical, the pineal gland controls the oscillation of brain activity between physical awareness and the non-physical dream state.

Base-12 Prime Structure of Red Blood Cells and the Circulatory System

In the same way the structure of the human brain and function of the mind seem to emerge as a base-12 prime fractal of 10^{11} neurons, our circulatory system of blood flow and oxygenation appears to be the resulting fractal of 10^{11} red blood cells. This is the approximate number of red blood cells we are each born with, this number increasing a hundredfold by the time we reach adulthood. It is therefore a beginning red blood count of about 10^{11} at which the human body achieves fractal independence as a living breathing entity just as a 10^{11} neuron count achieves a fractal consciousness of self.

The primary function of red blood cells is to absorb oxygen from the air we breathe, deliver it throughout the body and then remove carbon dioxide as a waste product to exhale. All vertebrates including mammals use red blood cells for this purpose as these cells are produced by bone marrow. Not only does the recirculating dynamic of blood from and to the lungs mimic the toroidal flow of the prime waveform, so too does the shape of the red blood cell itself. Referred to as a biconcave discocyte (just a fancy name for a torus, not a couple of caves converted into a dance club), the red blood cell is yet another embodiment of the base-12 prime waveform in 3D form.

Also, the red blood cells of mammals including humans lose their nuclei in a process called enucleation once the cells mature. This enables the cell to contain more oxygen-carrying hemoglobin. In fact, it is the only type of cell in our bodies which has this unique distinction. What's remarkable is how this mimics the way in which the nucleus or "black hole" of the prime waveform is a feature only of a mature fractal of the waveform; a feature which becomes hidden from physicality. This very same "enucleation" of galaxies occurs when the 10^{11} stellar density triggers formation of a central black hole.

Position 3 as the Base-12 Incubator of Life

Another mystery possibly solved by the base-12 prime theory is why the scale of complex life, of living cells, occurs right at the geometric mean between the smallest and largest fundamental scales of the universe; the Planck length at 10^{-35} metres and the Hubble length at 10^{26}.[44] The geometric mean of a series of n numbers is the nth root of the product of those numbers and represents the central number in a geometric progression. The Planck length is the smallest size theoretically possible while the Hubble length is the size of the observable universe. At these two extremes the universe appears to be at its simplest and between them its most complex.

The twelve positions of the prime waveform follow a logarithmic power scale like the spin cycle discussed earlier. Unlike spin, however, which follows a power of 10 scale from 10^0 to 10^{11} (as repeating full cycles of 10 in base-12), position relates to a single cycle of 12 and therefore follows powers of 12. Because position is the horizontal coordinate of each point on the base-12 circle, which is six positions wide, the position power scale increases from 12^0 at position 0 to 12^6 at position 6 and then decreases back again to 12^0 at position 12. As such, each universe life cycle from positions 0 through 6 goes from 12^0 to 12^6. The geometric mean for our universe life cycle would therefore be given by the calculation:

$$\sqrt[7]{12^0 \times 12^1 \times 12^2 \times 12^3 \times 12^4 \times 12^5 \times 12^6} = \sqrt[7]{12^{21}} = 12^{\frac{21}{7}} = 12^3$$

We see that position 3 is not only the very middle of the base-12 life cycle we perceive as the physical universe but also its geometric mean of 12^3. It is the stage at which space stops expanding, has the greatest probability window for diversity and complexity, the most direct and stable gravity and the only stage neutral in polarity for neutral atoms to form. All of these conditions make position 3 the ultimate "Goldilocks zone" most conducive for life. According to our base-12 prime model then, the scale of life occurs exactly where it should – at the maximum complexity of position 3 between the maximum simplicity of 0 and 6.

Perhaps it's no coincidence too that the Earth happens to be the third planet from the sun, at just the right distance for the liquid water and atmosphere necessary for biological life to thrive – just as it's no coincidence that the meaning of the number 3 in ancient numerology is that of the *catalyst*, that which creates change without changing itself.

The Mystery of the Number 137

While we're on the topic of why life occurs at the scale it does, let's take a little diversion and consider the enigmatic number 137. The number 137 is one of the most baffling numbers in science, appearing in many key places but without explanation.

The highest profile appearance of 137 is in the fine-structure constant (α = 1/137.036...) which characterizes the strength of the electromagnetic interaction between charged particles. This is one of the key physical constants of the universe. Other notable appearances of 137 include the current estimated age of the universe at 13.7 billion years old, the chlorophyll molecule critical to photosynthesis having 137 atoms and the 137.5-degree *golden angle* spacing that plants exhibit between branches, leaves and petals. Although these may appear to be just numerological coincidences, the base-12 prime waveform suggests otherwise. Let's examine a few of these to perhaps decipher the underlying significance of 137 in nature.

The fine-structure constant α is a dimensionless quantity related to the elementary charge *e* which is the unit of electric charge carried by the proton, the opposite polarity of which is the negative unit charge of the electron.[45] As such, *e* represents the strength of the electromagnetic interaction. Using a dimensionless value of *e* = 1 for the elementary charge to remain consistent with the dimensionless nature of the base-12 prime waveform, the fine-structure constant is related to *e* by the equation $\alpha = 2\pi e^2/hc$, where *h* is the Planck constant and *c* the speed of light.

The Planck constant h is the quantum of the electromagnetic interaction in which the energy E of a photon is directly proportional to its frequency f according to the relationship $h = E/f$. From our earlier discussion regarding the speed of light we saw that the energy of a photon in terms of the base-12 prime waveform is 2×12^6 and its frequency is 1. Therefore, the Planck constant can be expressed as $h = 2 \times 12^6$ which is indeed the wavelength of the base-12 prime waveform. This says that the 12^6 kinetic energy of the photon, the speed of light, only represents half the electromagnetic interaction and half the vibrational expression of the base-12 prime cycle. This points to the necessary existence of the other polarized half of the waveform and further confirmation of the universal matter/anti-matter duality upon which reality is based. Substituting $\pi = 12^6$ as representing a 180-degree rotation of the base-12 cycle, $h = 2 \times 12^6$ as derived above and $c = 12^6$, the equation for the fine-structure constant of $\alpha = 2\pi e^2/hc$ simplifies to a value of $2(12^6)(1)^2/((2 \times 12^6)12^6) = 1/12^6$ in terms of the base-12 prime waveform.

As the established estimated value of the fine-structure constant is 1/137 we discover a direct relationship between the strength of the electromagnetic interaction and the full 12^6 Higgs potential of the base-12 prime waveform as a probability space. This would agree with our earlier conjecture that the Higgs potential would be fully liberated as kinetic energy at position 0 as a massless photon travelling at the 12^6 speed of light and fully constrained at position 6 as the 12^6 potential energy of the Higgs rest state. This follows from the fact the 7-to-11 probability plane at positions 0 and 6 is horizontal with zero projection into our frame of reference and therefore zero probability of imparting mass.

In contrast, at position 3, the 7-to-11 probability plane is perpendicular and fully projected, enabling the full potential of the Higgs field to be imparted as mass. As we will see shortly, the probability of the top quark at position 3 is also precisely equal to 12^6. The heavyweight of the subatomic world, the top quark therefore represents the maximum physical manifestation of the base-12 prime waveform. This brings us to the golden ratio and photosynthesis.

The golden ratio or *phi* (φ) is nature's signature proportion of fractal growth and efficiency.[46] It represents the most efficient way for each successive generation to retain the defining characteristics of the one before, such that the child (a+b) is to the parent (a) as the parent (a) is to the grandparent (b), or (a+b)/a = a/b. In other words, each offspring represents the sum of the two before.

In its simplest geometric form, the golden ratio represents the ratio between the length of two portions of a line such that the entire length of the line divided by the long portion is equal to the long portion divided by the short portion. The golden angle is the radial expression of that same relationship as the ratio between two portions of a circle such that the entire circle (360 degrees) divided by the longer arc (222.5 degrees) is equal to the longer arc divided by the short arc (137.5 degrees), both equaling the golden ratio.

In practical terms, this 137.5 degree spacing from one branch, stem or leaf to the next ensures that plants have minimal overlap of appendages for maximum exposure to the Sun and an even distribution of weight of its limbs about its central stalk. So, just as 137 is the optimal energetic state (position 3) at which the 3D helical spin of the base-12 prime waveform is fully expressed as a 2D projection, it is also the optimal radial state at which the 3D helical structure of plants are best exposed to the sun, as illustrated below.

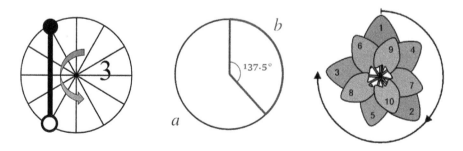

Position 3 as the Golden Angle of Maximum Exposure

Plants seek maximum exposure because they require sunlight for photosynthesis to convert carbon dioxide and water into glucose as food for the plant, with oxygen and water released as by-products.[47] The photosynthesis process can be summarized as follows:

$$6CO_2 + 12H_2O + \text{Light Energy} \rightarrow C_6H_{12}O_6 + 6O_2 + 6H_2O.$$

Notice how this process seems to reflect the 2:1 probability geometry of position 2 (12 water molecules and 6 carbon dioxide) being energized (by sunlight) to the balanced and fully projected 1:1 geometry of position 3 (6 oxygen and 6 water). Also, the molecule serving as solar collector for photosynthesis is chlorophyll a, $C_{55}H_{72}O_5N_4Mg$, consisting of 137 atoms. Thus, in the same way that the probability geometry of position 3 harnesses the full energetic potential of the base-12 prime cycle, chlorophyll harnesses the full energetic potential of photosynthesis.

So to recap, 137 is a measure of the interaction between matter and light, the golden angle that nature uses to optimize exposure to light and the molecular structure of the chlorophyll molecule which captures that light. These are all states of *maximum expression* in the physical world just as position 3 represents the maximum geometric expression of the prime waveform. So if position 6 is love as potential, position 3 is love fully expressed and the number 137 is its calling card.

A Golden Ratio Connection between Evolution and the Base-12 Prime Cycle

Not only can the golden ratio be described geometrically in terms of lines and angles but also mathematically by the Fibonacci series. This famous series consists of an infinite sequence of numbers such that each number is the sum of the two preceding numbers, starting from 0 and 1. This creates the sequence beginning with 0, 1, 1, 2, 3, 5, 8, 13, 21… in which the golden ratio equals the infinite limit of the ratio between two successive Fibonacci numbers. That limit is given by ½ +

(*sqrt* 5)/2 which equals the irrational number 1.618034… or, as a ratio, 1.618034 to 1.

The golden ratio pattern appears frequently in nature beyond just the arrangement of plant appendages as described above; we see it in the spiral form of sunflowers, sea shells and galaxies, the relative lengths and proportions of the human body and even the dimensions of the double helix spiral of DNA itself. The prevalence of this naturally occurring and aesthetically pleasing proportion has made it the subject of much attention and admiration over the centuries in art, architecture and science. Still, why it appears so often in nature as an evolutionary pattern has remained a mystery. When we examine the golden ratio more closely through the lens of the base-12 prime cycle, however, we discover what appears to be a direct mathematical link between prime numbers and the golden rule of nature. This link becomes apparent when we compare the geometry of the golden ratio to the base-12 prime cycle, depicted below.

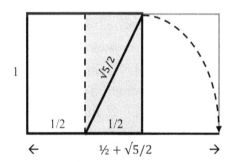 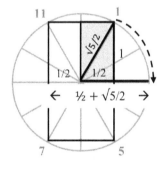

Golden Rectangle (left) versus Base-12 Prime Cycle (right)

For this little exercise, we will call upon the rectangular cousin of the golden ratio called – you guessed it – the golden rectangle, shown to the left. This is simply a rectangle with its width and height in golden ratio of (½ + (*sqrt* 5)/2) to 1. However, the way in which this rectangle can be constructed from a rotational feature of a square illustrates how the golden ratio naturally evolves from the base-12 cycle. To cut to the chase, where the square represents the first generation or *parents*, the resulting rectangle after the rotation is their golden ratio *child*.

We begin by drawing a 1 x 1 square representing the parents where, like the prime waveform, the vertical axis represents the imaginary dimension of unmanifested potential and the horizontal axis the real number line of manifested physicality. Next, we add a vertical line (dashed black) to divide that square in half. This vertical line creates two equal rectangles and represents the equal contribution of the father (on the left, unshaded) and mother (on the right, shaded) in the creation of a child – if you require further details, go ask your mother.

We then draw a diagonal line (solid black) from the lower left corner of mom's rectangle (i.e. where she and dad were, ahem, geometrically connected) to her upper right corner, forming a hypotenuse of length (*sqrt* 5)/2 (courtesy of the Pythagorean theorem) of a right-angle triangle with sides 1 and ½. Importantly, this child triangle is the simplest geometric shape that can be contained within the mother rectangle and which still retains her defining proportion of 2:1. Notice too how dad is no longer in the picture while mom is left carrying the load. Typical.

Finally, this hypotenuse of (*sqrt* 5)/2 when rotated down to the horizontal real axis (i.e. manifested into the world) creates the base of a new child rectangle which is in the golden ratio of ½ + (*sqrt* 5)/2 wide by 1 high. Thus, in a geometric and conceptual sense, the child (½ + (*sqrt* 5)/2) is in golden ratio to its parents (½ + ½ = 1) and represents the combination of ½ inherited from the father, ½ inherited from the mother and an additional (*sqrt* 5)/2 - ½ representing new adaptation (i.e. ½ father + ½ mother + (*sqrt* 5)/2 - ½ new = ½ + (*sqrt* 5)/2 child). Let's now relate this idea to the base-12 prime cycle to see if the same relationship holds from one generation or cycle to the next.

To the right of the golden rectangle is shown our base-12 prime cycle and the 4 x 2 rectangle inscribed inside by the four prime positions 1, 5, 7 and 11. We see right away that this 4 x 2 rectangle has the same 2:1 proportion as the 1 x ½ mother and father rectangles if we treat the height and width of the 4 x 2 rectangle as the positional differences between the four corners rather than physical lengths. In fact, each of the four prime positions relative to the origin of the prime cycle forms its own 1 x ½ rectangle with hypotenuse *(sqrt* 5)/2. But for our purposes, we will just focus on the position 1 rectangle as the golden ratio represents the relationship between a single unique or "prime" offspring and its parents.

Because the position 1 rectangle in the base-12 prime cycle is identical to the mother portion of the golden rectangle, the rotation of the hypotenuse from probability into reality results in the very same child geometry: ½ father + ½ mother + *(sqrt* 5)/2 - ½ new = ½ + *(sqrt* 5)/2 child. However, the prime cycle offers even further insight into the manifestation process when we consider the numerological rotation it follows.

First, we began at the very origin of the base-12 circle; a state of zero (vertical) probability and zero (horizontal) physicality between the two polarities of mother and father. The child is but a blink in their eyes – until the wine starts flowing, that is. The male and female polarities then "merge" from 2 into 1 and co-create a fertilized egg as the fundamental particle of life. It is at this moment of conception that the child's probability moves from the *potential* of position 0 to the *new beginnings* of position 1 as a new life within mom. That single cell then progresses to the *duality* of position 2, dividing again and again as the baby grows in tandem with mom.

Realizing his usefulness is now over, it is also around this time that dad becomes scarce. Dejected, he retreats to the shallow comforts of beer, pizza and sports on TV in order to watch other people score. And once the baby is ready it progresses to position 3 of the *catalyst* where, with a final push from mom, the baby is fully manifested into reality as a kicking and screaming bundle of joy.

It's also interesting to note from the discussion above how the new adaptive portion of the cycle of *(sqrt* 5)/2 - ½ = 0.618... represents the ability of each new generation to adapt to its environment with enhanced traits, beyond the ½ + ½ = 1 inherited from the previous generation. This suggests that if no new innovation occurs in a given cycle (i.e. no rotation of the hypotenuse to the horizontal reality axis) then the next generation will just be a repeat of the characteristics of the prior. In other words, no change.

We see this play out in nature with such species as the horseshoe crab. This living fossil hasn't evolved for millions of years because its ocean floor environment remained largely unchanged during that time. There was thus no energetic driver to activate the golden ratio of evolutionary change.

The Emergence of Love at The Quantum Scale

From the medium scale of life, we now zoom in to the tiny world of atoms and subatomic particles. As the quantum realm represents the smallest measurable scale of nature and the home of the most fundamental particles known, this is presumably the smallest fractal of the prime waveform we can physically explore. And as we will see, the over-achieving prime waveform will not disappoint.

The Standard Model of particle physics is the latest and greatest theory of quantum mechanics for describing the elementary particles of matter and the forces through which they interact.[48] This model also describes the anti-matter counterparts of the various matter particles, oppositely charged but otherwise identical. By all accounts, this model has proven tremendously successful in accurately predicting many observed aspects of nature at this teeny scale.

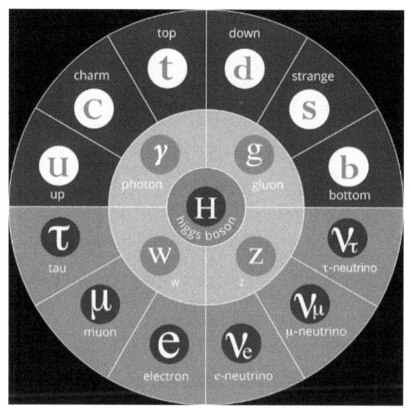

The Standard Model of Particle Physics
(*Source: https://atlas.cern/discover/physics*)

In the Standard Model, depicted above, there are twelve elementary matter particles called *fermions* and four force carriers called *gauge bosons*. The twelve fermions consist of six *quarks* (three with a +2/3 electric charge and three with a -1/3 charge) and six *leptons* (three with a -1 charge and three which are neutral with 0 charge). These are further grouped into three *generations* of particles based on related properties. Also, the way in which the quarks combine together follows a *three-colour rule*. All twelve fermions have a spin characteristic of ½ which relates to the angular momentum of the particle. Of those twelve fermions, only three are used to create all physical matter: the up quark, down quark and electron. The up and down quarks combine in combinations of three to form protons and neutrons which, in turn, combine in various ways to form the atoms of every known element in the periodic table.

The four types of gauge bosons are the *photon* as the carrier of the electromagnetic force, the *gluon* as carrier of the strong nuclear force and the *W* and *Z bosons* as carriers of the weak nuclear force. These four bosons all have a spin characteristic of 1. In addition, there's the recently confirmed *Higgs boson* which is the particle expression of the Higgs field that interacts with the other particles to give them their masses. The Higgs boson has a spin and charge of 0.

Right off the bat we see how remarkably aligned the subatomic world is to base-12 with everything structured according to 12 and its factors of 2, 3, 4 and 6. As we will discover, this structure emerges directly from the base-12 prime waveform. Still, despite the success of the Standard Model, it is not a complete story and leaves some pretty major gaps in explaining the quantum world, including the following head scratchers:

- Why are there twelve fermion particles and why are only three needed to make the atoms of all physical matter in the universe (the up quark, down quark and electron)?
- Why do those particles fall into two charge groupings of six quarks and six leptons and three generational groupings of four particles each?
- Why do the fermions and their force carrier bosons have the specific masses, electrical charges and spin characteristics they do?
- Why are there three forces (electromagnetism and the strong and weak nuclear interactions) which behave so differently at different scales and energies?
- Why isn't gravity addressed by the Standard Model and why is it so much weaker than the other forces at the subatomic scale?

The six types of quarks are referred to as the up, charm, top, down, strange and bottom. The first three possess +2/3 electrical charges while the latter three have charges of -1/3. In addition to the reappearance of the prime vibration's signature ratio of 2:1 in their relative charges of +2/3:-1/3, the quarks also appear to have a logical placement along our two prime waves if we were to give them positions based upon their relative characteristics. These characteristics include such defining properties as mass, electric charge, polarity, spin, predisposition to form into matter, relative stability and energetic potential.

With the help of the base-12 prime waveform as a probability blueprint, we will be able to decipher all of these particle properties simply based upon their positions (and those of their anti-particles) within the probability waveform. So let's get started.

A New Spin On Particle Spin

Before we begin assigning positions to the quarks and leptons, we need to understand their orientation within the figure-8 projection of the base-12 prime waveform. And to do that we need to better understand the characteristic referred to as spin.

In quantum mechanics, spin is the intrinsic angular momentum a particle possesses and represents the distinct quantum orientations it can have in each full rotation. Quarks and leptons all have a spin of ½ which means they have to complete two 360-degree rotations in order to return to their original state. The gauge bosons which include the photon, gluon and W and Z all have a spin of 1 while the Higgs boson has a spin of 0.

For the quarks and leptons to flip twice for a spin of ½, they would need to follow the inside of a figure-8 loop with a 180-degree twist in the middle. This would ensure that any particle position along the figure-8 path would always remain facing inward and towards the reality axis where matter manifests. Specifically, particles along the bottom boundaries would be oriented upward while particles along the upper boundaries would point downward. This figure-8 path is exactly what the prime waveform follows which means that all twelve fermions (and their twelve anti-particles) would need to have twenty-four positions directly on the figure-8 boundary of the prime waveform. As that's the total number of positions available along the upper and lower prime waves, we can expect to fully assign all the positions by the time we're done.

As the boson force carrier particles (photon, gluon, W, Z) represent energetic potentials between particle positions within the base-12 prime waveform and not particle positions themselves like the fermions, they are not subject to the twisting effect of the prime wave boundaries. This leaves them with one orientation only and thus a spin of 1.

Likewise, the only location within the prime waveform with the zero spin (and neutral charge) required of the Higgs boson is position 6, being the neutral intersection point between the two prime waves. This makes the Higgs boson the very heart of the base-12 prime waveform – the particle expression of love itself. In this light, its popularized nickname as the "God particle" takes on new meaning.

Particle Positions of the Six Quarks

At the atomic fractal of scale both half-cycles of the base-12 prime waveform are perceivable from our outside frame of reference, unlike our limited insider view of just the upper left quadrant at the universal scale. We can therefore perceive the full toroidal shape of atomic structure as expressed by the upper left and lower right quadrants of matter. All six quarks would therefore fall somewhere along the solid black probability boundary of matter while each of their anti-particles would be on its matching position on the dashed grey boundary of anti-matter.

The +2/3 charge of the up, charm and top quarks indicates they should be in the upper left quadrant matter region of positive charge while the three -1/3 negatively charged down, strange and bottom quarks should be in the negatively charged matter region of the lower right quadrant. Right away we can rule out positions 0, 6 and 12 for quarks because the probability window for each of those positions do not engage the reality axis. Other positions along the black wave we can rule out include positions 4, 5, <u>10</u> and <u>11</u> as the probability geometry at these positions are biased towards anti-matter rather than matter.

As the top quark is known to be heaviest in terms of mass and the least stable, it would be at the fully projected geometry of position 3 while the bottom quark would be at its polarized counterpart of position 9 six positions away. Next, the up and down quarks being the lightest and most stable places the up quark at position 1 and the down quark six positions away at position 7. Lastly, the charm and strange quarks having mass and stability between the top/bottom and up/down places them at positions 2 and 8, respectively. These placements of the three quark pairs of up/down, charm/strange and top/bottom therefore reveal the three generations of quarks to be based upon the polarized symmetry of each pair's probability profiles within the waveform.

Adding these six quark and six anti-quark placements to the base-12 prime waveform, we obtain the diagram below. Also shown are the Higgs boson at position 6 as well as the homes of the six leptons and their anti-particles to be explained next. For notation purposes, the upper particles of matter are capitalized in black font and their lower particle partners in lowercase black font. All corresponding anti-particles are shown in grey font with a minus sign and the Higgs boson by an H within a little heart (Aww!).

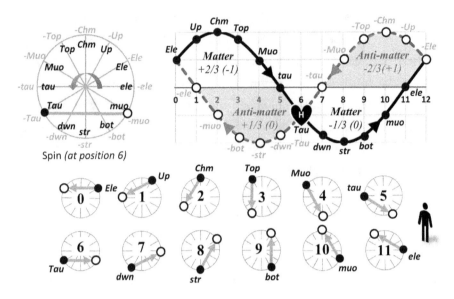

The Subatomic Fractal of Love

Note how the symmetrically opposite placement of each quark twelve positions away from its anti-quark along the prime waveform boundary places them at the *same* position on the base-12 unit circle. The same holds true for the leptons. This, again, is because the figure-8 boundary of the base-12 prime waveform represents two laps of the base-12 circle, a forward lap along the black prime wave of matter and a reverse lap along the dashed grey prime wave of anti-matter. This helps visualize the necessary duality of matter and anti-matter being created equally and why, if anti-matter encounters anti-matter, they annihilate on contact. Hey, sometimes love hurts.

Particle Positions of the Six Leptons

We will now apply the same mapping approach to the other six fermion particles, the leptons. Like the quarks, the leptons are grouped into three generations of related pairs: the electron and electron neutrino, muon and muon neutrino and tau and tau neutrino. However, the electric plane of the leptons is rotated 90-degrees relative to the magnetic plane of the quarks evidenced by the fact that a changing magnetic field creates an electric field perpendicular to itself but otherwise remaining in phase. In the diagram above, the electric plane would therefore be rotated 90 degrees backward into the page – essentially a top view of the magnetic plane as shown.

Unlike the upper and lower quarks having the charge ratio of +2/3:-1/3 of the magnetic plane, the upper and lower leptons are subject to the -1:0 charge ratio of the electric plane. Also, unlike the magnetic plane in which positions 0, 6 and 12 are not viable locations for quark particles, they are valid for leptons because the 90-degree rotation of the electric plane swings these positions through the reality axis. Positions 4, 5, 10 and 11 also become valid now that geometry of their probability windows is biased towards matter instead of anti-matter by virtue of the 90-degree rotation.

With this in mind, the six leptons would fall on the solid black probability boundary of matter while each of their anti-particles would be twelve positions away on the dashed grey boundary. The -1 charge of the electron, muon and tau places them in the upper left quadrant of the (90-degree rotated) electric plane while their three neutral neutrinos would be in the lower right quadrant.

In the upper left quadrant, positions 1, 2 and 3 are already occupied by the up, charm and top quarks. Assuming every unique position along the probability boundary must represent a unique particle (per the Pauli Exclusion principle which says that two particles cannot occupy the same quantum state), this leaves positions 0, 4 and 6 available for the electron, muon and tau.[49]

The electron would be at position 0 as it the lightest and most stable lepton requiring it to be furthest from the mass imparting effect of the Higgs at position 6. Its lower lepton partner, the electron neutrino, would likewise be at position 11 being the furthest (lightest) neutral position away from 6. As the probability profile for position 0 does not engage the reality axis, the electron would not form part of atomic nuclei but rather remain as a separate particle. The same applies to its anti-particle, the positron at position 12. Next, the tau being the heaviest and most unstable would be at position 6 closest to the Higgs, while its lower partner the tau neutrino would be at position 5 as the heaviest neutral position. Lastly, the muon being of intermediate mass and instability would be at position 4 with its muon neutrino at position 10.

As all twenty four available positions of the prime waveform are now fully occupied, we can conclude that the twelve quarks and leptons (and their anti-particles) identified thus far in the Standard Model likely accounts for all of them. This intuitively makes sense as we should expect no inefficiency or redundancy in nature; everything should have a purpose and be fully utilized. Once again the base-12 prime waveform model translates a rather obscure theoretical concept into a clear rational picture – of nature at its simplest and most efficient.

Particle Positions of the Five Bosons

Having assigned positions to all the fermion particles we will now characterize their interactions with each other and the particle expressions of those interactions, the bosons.

There are five bosons identified in the Standard Model according to the forces they appear to carry: the *photon* as the carrier of the electromagnetic force for all charged particles; the *W* and *Z* bosons as the carriers of the weak nuclear force; the *gluon* as the carrier of the strong nuclear force that keeps quarks together as protons and neutrons and those composite particles together in atomic nuclei, and the *Higgs boson* as the particle expression of the Higgs field which imparts mass to the other particles. All of the bosons have mass except the photon and gluon which travel freely through the Higgs field at the maximum speed limit of the universe, the speed of light.

As the simple vibrational structure of the prime waveform suggests the electromagnetic force to be the only force, the W, Z and gluon would all be higher energy expressions of the photon with the weak and strong nuclear forces their localized interactions at those energies. We will explore these particles next starting with the recently unveiled celebrity of the bunch, the Higgs boson.

The Higgs Boson as the Heart of the Waveform

The Higgs boson is the mass carrier particle for the Higgs field, the field which imparts mass to all the other particles and their anti-particles. More accurately, it is a quantized expression of the Higgs field that generates mass through its interaction with the other particles. It is also the only known particle with strictly mass but no charge or spin (i.e. angular momentum).

To have no spin and to impart mass equally to both halves of the base-12 prime waveform, the Higgs boson must be located at the very center of the cycle at position 6. This position represents the intersection of the two prime waves, a location of polarity reversal which would explain why the Higgs boson has no charge. Conceptually then, the Higgs boson represents the energetic potential between the two prime waves at the polarity reversal and origin of position 6. As this potential is the energy required to go from one prime wave to the other, there would presumably be a polarized opposite to the Higgs boson to balance out the process. This implies the existence of an anti-Higgs boson, also neutral in charge and with zero spin like its physical twin, though yet to be observed.

Note too how position 6 is offset below the reality axis where matter forms, essentially hidden within the "black hole" between positions 5 and 7 where matter cannot exist. This means that for the Higgs field to manifest as a particle it would need to acquire enough energy to reach the reality axis above. And that would be a lot of energy as evidenced by the hefty 125 GeV/c^2 mass measured for the Higgs and the very powerful and very expensive Large Hadron Collider at CERN in Geneva needed to coax it into making an appearance.

We saw earlier when taking a walk through the life cycle of the universe that it is this very offset of the polarity flip below position 6 that induces the charge ratio of +2:-1 for the polarization stage (epoch 2). In turn, it is the 120-degree phase difference between the two prime waves which creates this offset geometry to begin with. As such, the offset location of the Higgs boson and the ratio of the quark charges appears to be a direct consequence of the phase displacement between the two base-12 prime sine waves. This polarity offset I believe is the essential "symmetry breaking" mechanism that enables particles to acquire mass at the very beginning of each manifestation cycle.

Also, the Standard Model predicts that the Higgs boson is most likely to decay into a bottom quark pair, a decay mode experimentally observed in 2018.[50] We see from the Higgs position at the polarity flip of 6 that this would indeed be the case according to the base-12 probability profile. As the flow of the Higgs field from position 6 would move outward in both directions from below, the field would impart mass outward along the lower boundaries of the two prime waves. The first particles it would encounter within the Higgs' own mass range would indeed be the bottom quark at position 9 and anti-bottom at position 3 in the opposite direction.

Another intriguing clue that the Higgs field does indeed travel according to this outward "W" shaped probability path from the underside of position 6 is that this matches the curious "Mexican hat" shaped potential well that the Higgs field is known to exhibit.[51]

The Photon as the Figure-8 Boundary

As the carrier of the electromagnetic force for all electrically charged particles, the photon acts in both the magnetic plane of the quarks and electric plane of the leptons. The only fermions which do not interact with the photon are the three neutrinos and their anti-particles as they have no charge. Like the gluon, the massless photon travels at the speed of light.

Being massless, the photon doesn't interact with the mass-imparting influence of the Higgs field. As the mass acquired by a particle through the Higgs field is determined by the probability window between the upper and lower prime waves, the massless photon must therefore represent the outside boundary of the base-12 prime waveform only revealing itself as a "particle" of light whenever the waveform boundary intersects the reality axis. It is this strobe-light effect of photons appearing at seemingly discrete quantum positions that I suspect gives the illusion of wave-particle duality although light may actually be a continuous waveform in 4D space crisscrossing our 3D reality intermittently.

As prime positions 1, 5, 7 and 11 are the only positions at which the probability boundary intersects the neutral reality axis, it is only at these four locations where the photon would reveal itself as a particle of light. At sufficiently low energies as present in the cool universe today or in the electron orbitals of atoms it would be at positions 1 and 11 at which this would occur, while in the high energies near the polarity reversal of the newborn universe, at the event horizon of galactic black holes or in atomic nuclei it would be at positions 5 and 7.

In a similar way, every other probability position can become a tangible particle if energetically forced onto the reality axis. However, unlike the massless photon (and gluon as we will see next) which inherently crosses the reality axis at full speed on its own, all other positions need to be slowed down through the acquisition of mass by the Higgs field.

The Gluon as a High Energy Photon

The gluon is the exchange particle of the strong nuclear force which binds up and down quarks together into protons and neutrons and keeps those composite particles together within atomic nuclei. Only interacting with quarks, the gluon acts strictly in the magnetic plane. Our placement of the up and down quarks (and their anti-particles) on either side of central position 6 is consistent with this idea of the gluon binding a central nucleus. Again, being massless like the photon, the gluon travels unencumbered at the full speed of light.

The strong force is known to operate at two levels of intensity; full strength to bind individual quarks together (called the *strong interaction*) and a weaker level to hold protons and neutrons together (the *residual strong force*). The strong interaction between quarks has a further property called *asymptotic freedom*.[52] This is a weakening of the interaction at even higher energies and closer distances such that the quarks become freer to move. This is why in the extremely hot and dense conditions immediately after the supposed Big Bang, the universe would have contained free quarks not yet bound within protons and neutrons. It was only when the universe sufficiently grew in size and dropped in temperature that the full effect of the strong interaction would have kicked in to lock quarks together. As such, the three quarks contained within each proton and neutron are free to move relative to each other until their confinement distance is reached; sort of like marbles in a bag, the marbles can move freely inside the bag but need a lot of force to break through.

Based on the above, I believe the gluon represents the portion of the prime waveform between positions 5 and 7. This makes the nucleus the "black hole" of the atomic structure, if you will. We see from the geometry of positions 5 and 7 that the outer perimeter or "event horizon" of this region is highly polarized due to the opposite polarity and close proximity of those two positions. This would explain the powerful attractive force of the strong nuclear force to keep quarks together within this region. However, this attractive force would quickly fall away the closer the quarks are to neutral position 6 at the center of the nucleus. This would likewise explain asymptotic freedom and why gluons have never been found in isolation. We will explore the geometry of these interactions in greater depth shortly.

So despite the photon and gluon having different interaction strengths and ranges of influence, they are otherwise identical (i.e. massless, zero charge, spin 1). This all seems to agree with the prediction of the base-12 prime theory that the gluon is simply the photon's behavior within the intense region of the nucleus rather than a separate particle subject to a separate force.

The W Boson as a Six-Position Rotation

The W and Z bosons are the mediators of the weak interaction and, unlike the massless photon and gluon, do have mass. Another difference is that the W boson only interacts with left-handed fermions whereas the photon and gluon interact with fermions of either spin direction. The W boson comes in two types, the -1 charged W^- and its antiparticle the +1 charged W^+. The Z^0 on the other hand is neutral in charge and can only exchange energy between particles.

The W^- and W^+ can convert fermions or their anti-particles from one type to another by removing or adding a unit of electric charge, provided the pair of particles come from the same generation. The W boson can also decay into a particle/anti-particle pair or be produced by the combination thereof.

The first process is the mechanism behind the radioactive beta decay of a neutron into a proton in which a heavier down quark (-1/3 charge) transforms into a lighter up quark (+2/3 charge) by emitting a W^- boson (-1 charge). The W^- boson emitted in this process in turn emits a -1 charged electron and neutral anti-neutrino to complete the decay process (i.e. -1 W^- = -1 electron + 0 anti-neutrino). All heavier and therefore more unstable particles naturally seek to decay in this manner to lighter more stable states. In fact, without the help of the W boson, the heavier particles created in the early universe couldn't lose mass to form the light and stable up quark, down quark and electron needed for atoms to exist.

As a 180-degree counter clockwise rotation of the base-12 circle in the complex plane is equivalent to multiplying by -1 (i.e. i^2 = -1), the -1 charge of the W^- is equivalent to a 180-degree counter clockwise rotation of the base-12 circle and its emission a clockwise rotation. The W^+ is simply the reverse of this process, changing the up quark into a down quark through the absorption of a -1 charge. In terms of the base-12 prime waveform then, this 180-degree rotation equates to a six-position translation around the figure-8 probability boundary of the prime wave which agrees with our earlier placement of the fermions.

The Z Boson as a Twelve-Position Rotation

The neutral Z^0 boson decays into a particle-antiparticle pair of the same type of fermion or is created when such a pair annihilates, in contrast to the W boson which creates particle-antiparticle pairs of different particles. Because of the considerable mass of the Z^0 boson, only the heavier top/anti-top quarks are beyond its reach. Its heft also makes it very unstable and short-lived. In terms of probability geometry, the Z^0 boson represents a full 360-degree rotation of the base-12 unit circle or a twelve-position movement around the figure-8 probability boundary. For example, the annihilation of the up quark at position 1 with the anti-up quark at position <u>11</u> would emit a Z^0.

In summary, the prime waveform suggests that all five bosons – the photon, gluon, W, Z and Higgs – may all simply be particle expressions of the same electromagnetic (Higgs) field. This suggests the gluon is a high energy photon deriving its unique characteristics from the probability geometry between positions 5 and 7 of the base-12 prime waveform whereas the "conventional" photon applies to the lower energy portions of the waveform. The W boson, in turn, represents a six-position half rotation of the base-12 cycle with the W^+ and W^- representing opposite directions of that rotation, while the Z boson represents a full twelve-position rotation.

Also, as the Higgs boson is the mass expression of the Higgs field's potential before imparting mass to any other particles, it would follow that the probability of the other boson particles it works through would fully utilize that Higgs potential. As the only other bosons with mass are the W and Z (the photon and gluon being massless), the probability of the Higgs boson should hypothetically equal the sum of the probabilities of the W and Z which indeed they do.

The Crystalline Geometry of Electromagnetism

Based on the previous description of the gluon, the probability region between positions 5 and 7 of the prime waveform represents the atomic nucleus, literally the "black hole" and heart of the atom. The close proximity and opposite polarity of positions 5 and 7 means that the outer perimeter or "event horizon" of this region would be highly polarized, creating a strong containment region around the nucleus in which quarks would be securely held. However, the force of this containment would quickly fall the closer the quarks are to neutral position 6 at the center of the region. This inward weakening effect would explain the phenomenon called *asymptotic freedom*. Beyond positions 5 and 7 would likewise be an outward weakening effect represented by the geometry of a twelve-pointed star.

The 12D crystalline structure presented earlier on page 39 helps us visualize this idea graphically. Recall that this is the structure which becomes apparent when all twelve spin orientations of the 7-to-11 probability plane are viewed together as a complete cycle, essentially a cross-sectional view of the prime waveform. Labelling this diagram for the various interactions of the five bosons yields the image below.

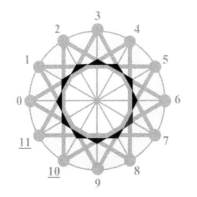

Strong Interaction (inner dark grey circle)

Weak Interaction (black triangular intersections between pairs of particles 6 positions away)

"Conventional" Electromagnetic interaction (outer light grey ring)

Geometry of the Strong, Weak and Conventional
Electromagnetic Interactions

Here we see that the rotation of the 7-to-11 probability plane through the full base-12 cycle (the twelve dark grey "barbells") creates a closed circular boundary with a diameter two positions wide, a circular probability boundary in 2D that would manifest as a spherical energetic boundary in our 3D space. Within this shell is a void region (unshaded) that would be asymptotically free of the influence of the probability plane rotating around it. This confinement region I suggest defines the strong interaction of the gluon which holds quarks together within protons and neutrons. Beyond this region, the influence of the star-shaped probability structure would decrease rapidly according to the tapered geometry of the twelve points of that star until reaching zero at the outer perimeter of the base-12 circle.

Next, recall that the weak interaction of the W boson represents a 180-degree rotation of six positions around the base-12 circle while the Z boson represents a full rotation of twelve positions. The most direct way to traverse the crystalline geometry from one side to the other (and back, in the case of the Z boson) without passing through the inner region of the strong interaction would be to follow the V-shaped path of the two barbells connecting those points. Using positions 1 and 7 of the up and down quarks to illustrate, this would mean crossing the 1-to-5 plane of the up quark until it intersects the 3-to-7 plane of the down quark, and then completing the path to 7. The little black triangular area formed where they intersect I suggest is the range of the weak interaction.

As we will see next, the proton and neutron also occur six positions apart and therefore follow the same probability geometry as the weak interaction of the W boson. This suggests that the force keeping protons and neutrons together in atomic nuclei may actually be a type of weak interaction rather than a residual aspect of the strong interaction as currently believed.

The electromagnetic force would be represented by the four position wide 7-to-11 probability plane which creates the prime waveform itself, the twelve rotations of which are the twelve barbells illustrated on page 94. These twelve barbells form the geometry of a twelve-pointed star which extends from the inner circle of the strong interaction to the outer perimeter of the base-12 circle. This would explain the infinite range of the electromagnetic force and the steadily narrowing geometry of each point the gradual weakening of that force with distance.

Proton and Neutron as the First Generation of Composite Particles

Other questions potentially answered by the prime waveform are why protons and neutrons make up the nuclei of all physical matter, why the up and down quarks are the exclusive building blocks of those composite particles and why the hydrogen atom is first to form.

As previously discussed, position 3 and position 9 are the only probability profiles within the prime waveform which are balanced in terms of equal polarity above and below the neutral reality axis. It would therefore be at only these two positions where neutral composite structures such as atoms could form.

Also, all particle interactions must follow the figure-8 path of the probability boundary which is left-to-right along the solid black prime wave of matter and right-to-left along the dashed prime wave of anti-matter. So, just as the universe life cycle was described in terms of the figure-8 cycle originating from position 6, we can apply it in the same way to the developmental cycle of atomic structure. However, unlike the physical universe in which only the upper left quadrant from position 0 to 5 is perceptible (our being within that half-cycle), the lower right quadrant from position 7 to 11 is also perceptible in the atomic cycle to us as outside observers.

- Moving left-to-right from position 6 as the origin and ground state of the Higgs field, we begin at position 7 with the creation of the -1/3 charged down quark. As its probability profile is entirely below the neutral reality axis, the down quark has a probability ratio of 0:1. This particle must manifest into matter here as its probability boundary intersects the neutral reality axis as one of the prime positions of 1, 5, 7 and 11. It is because the up quark, down quark, electron neutrino and tau neutrino (and their anti-particles) occur at these four prime positions that they are the prime building blocks involved in the formation of the nuclei of matter and anti-matter. The remaining eight fermions/

anti-fermions therefore represent transitional and inherently unstable quantum states between the four stable prime states.
- The probability profile then changes to a 1:2 ratio at position 8 which prompts the formation of the neutral neutron from one +2/3 charged up quark (position 1 of the up quark being the only probability profile with the required 1:0 ratio and of opposite charge to that of the down quark) and two -1/3 charged down quarks (for a probability ratio of 0:2). Thus, the 1:0 ratio of the up quark must combine with the 0:2 ratio of two down quarks in order to achieve the 1:2 ratio dictated by position 8, the result of which is the neutron as a composite particle.
- The probability profile changes again at position 9 to a balanced 1:1 ratio requiring the addition of a unit of upper probability (1:0 ratio) that is also neutral in charge so as to maintain the already neutral stability of the neutron (1:2 ratio). This can only be met by the 1:0 probability profile of the electron neutrino at position 11. Thus, position 9 represents the neutron and electron neutrino together as a stable particle pair (1:2 + 1:0 = 2:2 or 1:1) though unbounded due to their individual neutrality.
- At position 10 the probability profile changes to 2:1 which is that of the +1 charged proton (two +2/3 charged up quarks and one -1/3 charged down quark). This requires the aforementioned beta decay process of replacing a down quark (0:1 ratio) with an up quark (1:0 ratio) in order to change the neutron/neutrino pair of position 9 (1:2 + 1:0 = 2:2 ratio) into the proton of position 10 (2:1 ratio).
- This ratio change from 2:2 of position 9 (neutron with neutrino) to 2:1 of position 10 (proton without neutrino) therefore requires the elimination of the 1:0 neutrino at position 11 (through the creation of its -1:0 anti-particle at position 1, annihilating both) and a further ratio change of +1:-1 which can only be met by the addition of the 1:0 up quark of position 1 and subtraction of the 0:1 down quark of position 7.

- As the replacement of the -1/3 charged down quark by the +2/3 charged up quark also requires the elimination of a -1 charge without impacting the overall probability ratio, the -1 charged electron is created at position 0 as a point probability. Because the probability profile of position 0 as a point probability does not engage the reality axis, the electron remains separate as a fundamental particle though still integral to the atomic structure.
- At position 1 the +2/3 charged up quark is created, its probability profile touching the reality axis as one of the four prime particles which must form.
- From the +2/3 charged up quark at position 1 (1:0 ratio) the probability profile changes at position 2 to a 2:1 ratio. This creates the +1 charged proton from two +2/3 charged up quarks and one -1/3 charged down quark (2 x 1:0 + 0:1 = 2:1).
- At position 3 the probability geometry then changes to a balanced and neutral 1:1 ratio which requires the combining of the -1 charge of the electron to the +1 charge of the proton to achieve that neutrality. As the unbalanced 2:1 ratio of the proton is unchanged by the addition of the point probability of the electron, the neutral tau neutrino (0:1) is also created at position 5 to achieve balance (2:1 + 0:1 = 2:2 ratio). This creates the hydrogen atom as the simplest and first atomic structure to form, along with the tau neutrino as a free particle.
- The probability ratio then changes from the 2:2 ratio of position 3 (proton with electron and tau neutrino) to the 1:2 ratio of position 4 (neutron alone) which essentially triggers the beta decay process in reverse; of replacing an up quark with a down quark.
- This requires the elimination of the 0:1 tau neutrino at position 5 (through the creation of its 0:-1 anti-particle at position 7, annihilating both) and a further ratio change of -1:+1 which can only be met by the addition of 0:1 down quark of position 7 and

subtraction of the 1:0 up quark of position 1. As the replacement of the +2/3 charged up quark by the -1/3 charged down quark also requires the absorption of a -1 charge without impacting the overall probability ratio, the +1 charged anti-tau is created at position 6 as a point probability but immediately annihilated by the -1 charged electron liberated from the hydrogen atom.

From the above description of the atomic cycle we see that the most stable quantum state in the right half of the cycle is achieved at the 1:1 probability profile of position 9. Again, this position represents the neutron (1:2 ratio) and electron neutrino (1:0 ratio) together as a stable pair (1:2 + 1:0 = 2:2). As the neutron and neutrino are already individually neutral in charge they do not need to combine into a composite particle to achieve the required neutrality of stable matter. Likewise, the most stable state in the left half of the cycle occurs at the 1:1 probability profile of position 3, that of the +1 charged proton (2:1 ratio), -1 charged electron (point probability) and neutral tau neutrino (0:1 ratio) together as a stable combination (2:1 + 0 + 0:1 = 2:2).

In short, positions 3 and 9 represent the prime vibration at its most manifested, stable and neutral and all other quantum positions play a supporting role. The right side of the cycle achieves that optimal expression through the neutron and electron neutrino and the left side through the hydrogen atom and tau neutrino. These are the neutral ingredients from which everything in the physical universe should theoretically be made according to the prime vibration, again agreeing with observation.

Note too that position 8 being six positions away from position 2 (i.e. a 180-degree rotation in the complex plane) makes the neutron the lower "composite particle partner" of the proton, just as the strange quark is the lower particle partner of the charm quark at those same positions. The proton and neutron therefore constitute a generational partnership specific to composite particles, beyond the established three for fundamental fermion particles. Further, the proton and neutron being the direct result of the base-12 prime waveform for matter implies that anti-protons and anti-neutrons would also be created in the anti-matter regions to maintain energetic balance and symmetry overall.

Still, we see that the proton at position 2 is four positions away from position 6 while the neutron at position 8 is only two positions away. As positions increase in energy and instability the closer they are to the Higgs potential at position 6, the neutron would be less stable and slightly heavier than the proton. This would drive the beta decay process described earlier of converting neutrons into protons via the W boson, particularly once the early universe cooled sufficiently, and would explain why hydrogen (consisting of a proton and electron) is the first stable atom to form.

Stellar Cycle as a Larger Fractal of the Atomic

If the base-12 prime waveform indeed operates at all scales, an intriguing implication of the proton/neutron cycle of the atom is that it should also apply to any larger self-contained composite of atoms such as stars. When we consider the typical life cycle of a star, this appears to be the case.

A star begins from a nebula cloud of loose gas and dust particles as characterized by positions 0 and 1 of the atomic cycle. As gravity intensifies from position 1 to 2, the cloud collapses and heats up as a protostar core consisting primarily of loose protons and electrons. As gravity increases further from position 2 to 3, the protons and electrons combine to form neutral hydrogen which kick starts the nuclear fusion of hydrogen into helium as fuel for the star.

Once the star runs out of fuel at position 3, it contracts at position 4 into a dense white dwarf star (of atoms still packed with electrons) or, if more massive, an even denser neutron star (with electrons and protons replaced by neutrons).

If massive enough, the neutron star collapses further to an event horizon at position 5 and into a black hole. As the black hole's intense gravity pulls inward while also spinning, it would exhibit a swirling glowing ring of high energy particles and anti-particles annihilating upon contact. This would produce a visible accretion disc around the black hole as finally observed in 2019. And once the black hole eventually runs out of steam, it disappears from view.

Particle Probability and Mass

In the Standard Model, the masses of particles are typically expressed in units of MeV/c^2 or mega-electron volts (MeV) divided by the square of the speed of light in a vacuum (c). An electron volt (eV) is the amount of energy an electron gains when accelerated (in m/s^2) by 1 volt of electricity. This follows from $E = mc^2$ in which the mass of a particle at rest is given by $m = E/c^2$. Since we cannot directly weigh such point particles of matter their masses must be derived indirectly from their probabilities.

Courtesy of the Schrödinger wave equation of quantum mechanics, the probability of a particle appearing in a certain place is equal to the squared amplitude of its probability wavefunction in the complex plane.[53] More specifically, it's the squared sum of the amplitudes of *all* possible ways the event can happen or the "sum of histories". That is, we need to sum together all the possible probabilities before squaring the total because we are adding together waves which can have positive or negative amplitudes. And since waves can interfere constructively or destructively depending on the relative signs of their amplitudes, adding them together beforehand properly reflects that interference effect to provide the net amplitude of the combined wave. This is the same addition we used to obtain our two prime waves from the base-12 cycle; we added together the 1 and 7 prime position waves to get one prime wave and added together the 5 and 11 waves to get the other.

Particle Probability as Mass Squared

Once I started putting numbers to the geometry of the particle positions just described and comparing them to the known masses for the particles, a very surprising relationship became apparent. The base-12 prime waveform seems to provide the probability of each particle directly such that the particle's mass simply equals the square root of its probability. This seems to eliminate the painstaking preliminary step of calculating all possible probabilities (or at least the significant ones) as normally required with the traditional "sum of histories" approach.

And it's not just the net probability that the geometry of the prime waveform seems to provide for each particle. It also reveals information about the particle's energetic polarity, spin, stability and its relationship to the other particles, again directly from the geometry of the waveform and without the need for involved math. From this we gain valuable insight into their charge relationships and proportional structure that's lost when only net probabilities are considered. And, as we will see shortly, this enables us to understand how all the particles work together in vibrational balance and *must* have the masses they do.

Based upon this insight, the mass of each upper fermion particle (i.e. within the upper left quadrant from position 0 to 5) is given by the square root of its probability, where the probability equals the product of the position's projected magnitude, position and spin (i.e. the product of its three quaternion coordinates as a 3D rotation in 4D space and defined by ijk = -1). Likewise, the mass of each lower fermion (i.e. within the lower right quadrant from position 6 to 11) is given by the square root of its probability except that probability this time equals the product of its upper particle partner's mass (square root of probability) times the lower particle's spin. We will now walk through these calculations to clarify how this works.

Converting Waveform Probability from $(eV/12^{12})^2$ to $(MeV/c^2)2$

In order to express particle mass in the conventional units of MeV/c^2, any upper particle probability derived from the prime waveform must first be multiplied by a conversion factor of 10^4. This is required because the base-12 prime waveform is based on a fixed energetic potential of 12^6 as a unitless constant rather than a speed of light in meters per second, so 12^6 must be multiplied by 10^2 to express that potential in meters per second. This recalibrated speed of light of c = $12^6 \times 10^2$ m/s enables us to express the conventional MeV/c^2 units of measure for particle mass as $MeV/(12^6 \times 10^2)^2$ or $MeV/(12^{12} \times 10^4)$. Further, we can safely assume the energy expressed by the base-12 prime waveform would be in the most fundamental units of electron-volts (eV) and not the usual convention of MeV or 10^6 electron-volts preferred by physicists used to mega-sized coffee mugs.

As such, our units of particle mass of $MeV/(12^{12} \times 10^4)$ would instead become $eV \times 10^6/(12^{12} \times 10^4) = eV \times 10^2/12^{12} = eV/12^{12} \times 10^2$. Since *mass*^{2} = *probability*, our units of probability would therefore be expressed as mass^2 = $(eV/12^{12} \times 10^2)^2 = eV^2/12^{24} \times 10^4$. Thus, any probability values we obtain from the prime waveform will need to be scaled up by a factor of 10^4 before using them to calculate particle mass in the conventional units of MeV/c^2. Note that this conversion factor would only be required for upper particle probabilities since lower particle probability as the square root of upper probability is derived from that already converted upper probability.

Projected Magnitude, Position and Spin for the Twelve Fermions

We will now determine the three probability factors of magnitude, position and spin for the fermions based upon the vertical geometry, lateral position and rotational orientation, respectively, of their probability profiles. All of these values are obtained directly from the base-12 prime waveform on page 94, the fractal of love at the quantum scale.

Magnitude Factors as Probability Geometries

The magnitude factor for each particle is derived from the vertical geometry of its probability profile as projected into the magnetic plane. As this geometry may fall above and/or below the neutral reality axis, the probability magnitudes are scaled according to the +2/3:-1/3 or +2:-1 charge ratio of the magnetic plane. The portion of any probability profile above the reality axis is therefore taken at full value while the portion below is taken at 50%. As lower particle probability is derived from the square root of upper particle probability, the magnitude factors of only the six upper fermions are required in determining the probabilities of all twelve fermions.

The electron at position 0 as a point probability of 0.25 above the reality axis gives it a magnitude factor of 0.25 while the up quark at position 1 being sin 60°/2 or 0.433 above the reality axis would have a magnitude factor of 0.433.

The charm quark at position 2 has a probability profile of 0.5 above the reality axis and 0.25 below. This 2:1 probability ratio also makes it a little special. As this is equivalent to a 180-degree rotation of the lower probability boundary and a further doubling of the resulting upper amplitude, the magnitude factor of the charm quark would be $((0.5 + 0.25/2) \times 12^2)^2 \times 2$ or 16,200. Note that the 122 position factor is included here in the squaring operation as the position factor and magnitude factor together determine the vertical amplitude of the projected probability and are therefore both subject to the 180-degree rotation.

The top quark at position 3 is also unique in that its probability profile is in a 1:1 ratio above and below the reality axis. This represents the probability plane fully projected and is equivalent to the upper probability boundary being rotated a full 180 degrees about the reality axis. As a 180-degree rotation in the complex plane squares the result, the magnitude/position factor of the top quark is $(1 \times 12^3)^2$ or 12^6, with the 12^3 position factor again included in the squaring operation.

The muon at position 4 with a probability profile of 0.25 above and 0.5 below would have a magnitude factor of 0.25 + 0.5/2 or 0.5. The tau at position 6 is a point probability like the electron at position 0 but differs in that position 6 is hidden below the reality axis requiring the tau to manifest at position 5. This gives the tau a point probability of 0.25 below the reality axis and thus a magnitude factor of 0.25/2 or 0.125. Note that of the electron, muon and tau only the muon possesses a lopsided +1:-2 probability geometry which might explain the odd "wobble" that this particle exhibits in experiments.

Position Factors as Powers of 12

The position factor is the power of 12 value of the base-12 position along the neutral reality axis at which the particle manifests. As depicted below, the position factor accounts for a particle's proximity to position 6 as the energetic center and source of the Higgs field, the "black hole" of the base-12 prime waveform. This is why the scale increases from 12^0 to 12^6 from positions 0 to 6 and then decreases from 12^6 to 12^0 from positions 6 to 12 – the closer a particle's probability profile is to the central Higgs at position 6, the greater the influence of the Higgs in imparting mass to that particle.

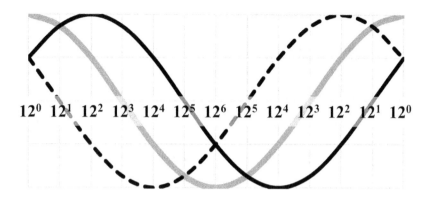

Position Factors as Powers of 12

This amplification of the position factor the closer it is to the center of the waveform agrees with the mass concentration of each atom, the nucleus and home of the hefty protons and neutrons, being at the very center of the atomic structure and the lightest charged particle, the electron, being furthest away at position 0. In fact it agrees with all the positions we previously assigned to the quarks, leptons and Higgs boson; positions we assigned simply based on matching each particle's known properties to the geometry of the available probability profiles.

As lower particle probability is derived from the square root of upper particle probability, only the position factors of the six upper fermions are involved in determining the probabilities of all twelve fermions. These position factors are listed below.

- Electron (position 0): 12^0 position factor
- Up quark (position 1): 12^1 position factor
- Charm quark (position 2): 12^2 position factor
- Top quark (position 3): 12^3 position factor
- Muon (position 4): 12^4 position factor
- Tau (position 5): 12^5 position factor (manifests at position 5 although the tau's probability profile is located at 6)

Spin Factors as Powers of 10

As with the position factor, the spin factor follows a power scale that increases to the mid-point of the base-12 cycle and then decreases again. However, unlike the position factor which increases the closer a particle position is to the mass center at 6, the spin factor increases in power the further a position rotates into the complex plane from our perspective.

The spin factor for each particle is given by the difference between the exponents of the two prime wave power terms at that particle position and represents the particle's angular momentum or *spin* into the extra dimension of space. This is the same spin characteristic introduced earlier as that caused by the twisted nature of the probability boundaries of the base-12 prime wave and which gives rise to the fermions' intrinsic spin of ½.

Each position around the unit circle as a power of 10 represents that position's angular momentum as a vector tangent to the circle and pointing in the direction of rotation (which is always counter clockwise on the complex unit circle). As such, adding those two vectors gives the net spin vector for the particle's probability plane.

The diagram below illustrates how that vector addition would work for the up quark. To add the 10^6 vector to the 10^2 vector, picture the end of the 10^2 vector (solid black arrow) connected to the start of the 10^6 vector (dashed black arrow). The new arrow (in grey) needed to complete the triangle is the net vector resulting from the addition. As the spin vector for any position must be tangent to the unit circle and pointing in a counter clockwise direction, we see that our grey net vector corresponds to position 10^4 or halfway between the two prime waves. As the net vector is angled downward in the negative direction at that location, the up quark would be assigned a negative power of 10^{-4} reflecting its negative spin.

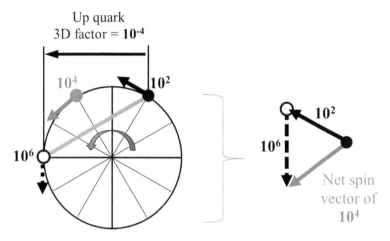

Spin Factor for Up Quark as a Vector in the Complex Plane

Again, it is the projected depth of the plane that determines the spin power factor for quarks because the spin power cycle in the magnetic plane behaves like a linear scale from 10^0 to 10^6 and back again. As each particle's probability is defined by the constant four-position wide plane created between the two twisting prime waves as probability boundaries, the magnitude of the 3D scaling factor is given by the projected depth as a power of 10 difference between those two prime waves.

The spin factor for each upper quark (up, charm, top) is determined this very same way, by the projected depth of its probability plane as a vector into the complex plane with the sign of its exponent (positive or negative) given by its direction (up or down). These spin factors are illustrated in the diagram below.

- Up quark: 10^{-4} spin factor
- Charm quark: 10^{-2} spin factor
- Top quark: 10^0 spin factor

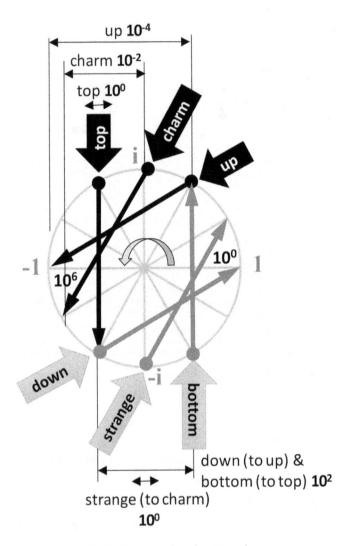

Spin Factors for the Quarks

The spin factors for the three lower quarks (down, strange, bottom) are determined slightly differently than their upper quark partners due to the 180-degree rotational relationship between the upper and lower quarks such that the lower probability equals the square root of upper probability. This links each lower quark to its upper quark partner such that the lower quark's spin factor is given by the projected depth between the two particle positions, effectively reversing the upper quark's individual spin factor and replacing it with its paired factor.

As the top quark has a zero spin factor (i.e. 10^0) there is nothing that needs reversing before obtaining the bottom quark's spin factor of 10^2 relative to the top quark. However, the down and strange quarks are affected. Because the up quark has a 10^{-4} spin factor on its own whereas the up/down quarks combined have a spin factor of 10^2 in the other direction, a reversal factor of 10^2 must be applied. Also, as the probability geometries of the up and down quarks are going from +1:0 and 0:-1 respectively to a combined geometry of +1:-1, that which is associated with a 180-degree rotation and a squaring of probability, the reversal factor of 10^2 likewise needs to be adjusted by its square root to 10^1.

In the case of the charm/strange quark pair, the charm quark on its own has a 10^{-2} spin factor while the charm/strange quarks combined have a spin factor of zero or 10^0. This therefore needs to be reversed by multiplying by 10^2. The probability geometry is also going from +2:-1 (the squaring and doubling of probability) for the charm quark on its own to +1:-1 (the squaring of probability only) in combination with the strange quark. Thus, the squaring remains in place for the strange quark and only the doubling needs to be reversed by halving the probability. The resultant spin factors of the lower quarks based on the above reasoning are listed below, all of which represent upward vectors (positive exponents):

- Down quark: 10^1 spin factor
- Strange quark: 10^0 spin factor
- Bottom quark: 10^2 spin factor

The spin factors for the three upper leptons (electron, muon, tau) are also projected depths like the upper quarks. However, as the electric plane of the leptons is 90-degrees rotated relative to the magnetic plane of the quarks, this depth is measured from a "top view" perspective of the magnetic plane. Also, because the charge ratio of the leptons is -1:0, the total charge is one-sided and fully applied to the upper leptons. Thus the height projected for the upper leptons is as point probabilities relative to the full height of the base-12 circle rather than as four-position vectors like the upper quarks. This results in the following spin factors for the upper leptons, as illustrated below:

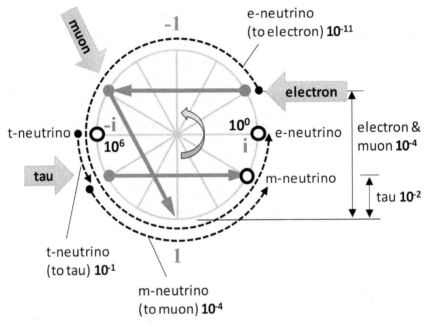

Spin Factors for the Leptons

- Electron: 10^{-4} spin factor
- Muon: 10^{-4} spin factor
- Tau: 10^{-2} spin factor

Because the spin power cycle in the electric plane behaves like a rotational cycle from 10^0 to 10^{11} rather than a linear half-cycle as in the magnetic plane, the spin factor for the three lower leptons (neutrinos) are based on their rotational positions around the base-12 circle relative to the probability planes of their upper lepton partners rather than their relative projected depth.

As the electron and tau have horizontal probability planes, their positions as spin vectors relative to their lower lepton partners are the same as if they were point probabilities. Thus, the electron neutrino at 10^0 relative to the electron at 10^{11} has a spin factor of 10^{-11} while the tau neutrino at 10^6 relative to the tau at 10^5 has a spin factor of 10^{-1}. The muon with an angled probability plane (from 10^7 to 10^3) would place its spin vector at 10^5. As the muon neutrino is at 10^1 its spin factor relative to the muon would therefore be $10^{1-5} = 10^{-4}$. This gives the following spin factors for the lower leptons:

- Electron neutrino: 10^{-11} spin factor
- Muon neutrino: 10^{-4} spin factor
- Tau neutrino: 10^{-1} spin factor

Why Are Neutrinos South Paws?

While on the subject of neutrinos, another mystery of particle physics possibly answered by the base-12 prime waveform is why neutrinos have a strictly clockwise or "left-handed" spin if looking in the direction of the particle moving towards us.

As radial positions, the relative position of each lower lepton is expressed in terms of the inherent counter clockwise rotation of the unit circle in the complex plane. As the base-12 unit circle as I have depicted it is in the direction of particles moving away from us and "into the page" (i.e. from the perspective of their source), its counter clockwise rotation would appear as a clockwise spin from our perspective as observers on the receiving end (i.e. from within the upper left quadrant, looking "back in time" towards the source).

Probabilities and Masses of the Twelve Fermions

With the magnitude, position and spin factors now determined for the twelve fermions, their probabilities and masses can now be calculated.

Each of the predicted values are in bold followed by its experimentally measured value revealing how well they agree.[54]

The six upper fermions follow the formula:

Probability = magnitude x position x spin x 10^4 conversion, where *mass = $\sqrt{Probability}$*

- Electron: Probability = $0.25 \times 12^0 \times 10^{-4} \times 10^4 = 0.25$ (MeV/c²)²
  Mass = $\sqrt{0.25}$ = **0.50 MeV/c²** *versus* **0.511 measured**
- Up quark: Probability = $\sin 60°/2 \times 12^1 \times 10^{-4} \times 10^4 = 5.20$ (MeV/c²)²
  Mass = $\sqrt{5.20}$ = **2.28 MeV/c²** *versus* **2.3 measured**
- Charm quark: Probability = $((0.5+0.25/2) \times 12^2)^2 \times 2 \times 10^{-2} \times 10^4$ = 1,620,000 (MeV/c²)²
  Mass = $\sqrt{1,620,000}$ = **1,273 MeV/c²** *versus* **1,275 measured**
- Top quark: Probability = $(1 \times 12^3)^2 \times 10^0 \times 10^4$ = 29,859,840,000 (MeV/c²)² Mass = $\sqrt{29,859,840,000}$ = **172,800 MeV/c²** *versus* **173,210 measured**
- Muon: Probability = $(0.25+0.5/2) \times 12^4 \times 10^{-4} \times 10^4 = 10,368$ (MeV/c²)²
  Mass = $\sqrt{10,368}$ = **101.8 MeV/c²** *versus* **105.7 measured**
- Tau: Probability = $0.25/2 \times 12^5 \times 10^{-2} \times 10^4 = 3,110,400$ (MeV/c²)²
  Mass = $\sqrt{3,110,400}$ = **1,764 MeV/c²** *versus* **1,777 measured**

The six lower fermions are derived from their upper particle partner with the formula:

$Probability_{lower}$ = $\sqrt{Probability_{upper}} \times spin_{lower}$, where *mass = $\sqrt{Probability}$*

- Electron neutrino: Probability = $\sqrt{0.25} \times 10^{-11} = 0.50 \times 10^{-11}$ (MeV/c²)²
  Mass = $\sqrt{0.50 \times 10^{-11}}$ = **2.2x10⁻⁶ MeV/c²** *versus* **<2x10⁻⁶ measured**
- Down quark: Probability = $\sqrt{5.20} \times 10^1 = 22.8$ (MeV/c²)²
  Mass = $\sqrt{22.8}$ = **4.77 MeV/c²** *versus* **4.8 measured**
- Strange quark: Probability = $\sqrt{1,620,000}/2 \times 10^2 \times 10^0$ = 9,000 (MeV/c²)²
  Mass = $\sqrt{9,000}$ = **94.9 MeV/c²** *versus* **95 measured**

- Bottom quark: Probability = $\sqrt{29{,}859{,}840{,}000}$ x 10^2 = 17,280,000 (MeV/c²)²
  Mass = $\sqrt{17{,}280{,}000}$ = **4,157 MeV/c²** *versus* **4,180 measured**
- Muon neutrino: Probability = $\sqrt{10{,}386}$ x 10^{-4} = 0.01 (MeV/c²)²
  Mass = $\sqrt{0.01}$ = **0.10 MeV/c²** *versus* **<0.19 measured**
- Tau neutrino: Probability = $\sqrt{3{,}110{,}400}$ x 10^{-1} = 176.4 (MeV/c²)²
  Mass = $\sqrt{176.4}$ = **13.3 MeV/c²** *versus* **<18.2 measured**

Probabilities and Masses of the W, Z and Higgs Bosons

The probabilities and masses for the three massive bosons, the W, Z and Higgs, are determined in the same manner as just described for the upper fermions. The only difference is that the bosons represent interactions between pairs of positions along the figure-8 probability boundary whereas the fermions represent the individual positions themselves. The photon and gluon (as a higher energy photon) are excluded from this discussion as they are massless.

Higgs Boson – Position 6 in Magnetic Plane

The probability geometry of the Higgs boson at position 6 is similar to that of the electron at position 0 in that both have a 0.25 magnitude relative to the neutral reality axis. However, where position 0 of the electron represents a point probability, position 6 of the Higgs is an intersection of the two prime waves and therefore an overlap of two point probabilities with twice the electron's probability magnitude. As position 6 is a polarity reversal, the Higgs boson would have no angular momentum and therefore a spin factor of 10^0. Also, the probability position factor for position 6 is 12^6. Utilizing the same formula as used for the upper fermions, the probability and mass of the Higgs boson is given by:

Probability = magnitude x position x spin x 10^4 conversion, where *Mass = $\sqrt{Probability}$*

- Probability$_{Higgs}$ = 0.25x2x12^6x10^0x10^4 = 14,929,920,000 (MeV/c^2)^2
  Mass = $\sqrt{14,929,920,000}$ = **122,188 MeV/c^2** *versus* **125,350 measured**

W Boson – Half Rotation in the Complex Plane

The probability geometry of the W boson is revealed through the beta decay process in which the neutron (consisting of one up quark and two down quarks) transforms into the proton (two up quarks and one down quark) via a -1/3 charged down quark changing into a +2/3 charged up quark by emitting a -1 charged W⁻ boson, which in turn emits a -1 charged electron and neutral anti-neutrino.

As position 7 of the down quark is six positions away from position 1 of the up quark, the emission of the -1 charge of the W⁻ boson is equivalent to a 180-degree counter clockwise rotation around the base-12 unit circle. The probability of the W⁻ is therefore represented by the area of the base-12 circle swept out by the combined geometry of the up quark and the down quark, each of which has an amplitude of sin(60)/2 relative to the neutral reality axis. The probability and mass of the W boson is therefore given by:

Probability = magnitude x position x spin x 10^4 conversion, where *Mass = $\sqrt{Probability}$*

- Probability$_W$ = sin(60)/2 x 12^6/2 x 10^0 x 10^4 = 6,464,844,998 (MeV/c^2)^2
  Mass = $\sqrt{6,464,844,998}$ = **80,404 MeV/c^2** *versus* **80,380 measured**

Z Boson – Full Rotation in the Complex Plane

The neutral Z boson is responsible for creating particle-antiparticle pairs of the same type of particle and is therefore equivalent to a full 360-degree rotation around the base-12 circle. Also, as the Higgs boson is the mass expression of the Higgs field's potential before imparting mass to any other particles, it would follow that the probability of the other boson particles it works through would fully utilize that Higgs potential. As the only other bosons with mass are the W and Z (the photon and gluon being massless), the probability of the Higgs boson should hypothetically equal the sum of the probabilities of the W and Z under this logic. Or stated another way, knowing the Higgs and W probabilities the Z should be given by their difference:

Probability $_Z$ = Probability $_{Higgs}$ − Probability $_W$ = 14,929,920,000 − 6,464,844,998 = 8,465,075,002 (MeV/c²)²
Mass = $\sqrt{8,465,070,000}$ = **92,006 MeV/c²** *versus* **91,190 measured**

Recap of Particle Probabilities and Masses

Everything we just walked through highlights the surprising but undeniable connection between the base-12 prime vibration as a probability blueprint of nature and the masses of the subatomic particles. Having covered a lot of ground to get to this point, let's do a quick recap to appreciate how it all works together as a highly efficient and unified system.

We began by placing each fermion particle along the probability waveform according to its known characteristics including whether it's a quark (on magnetic plane) or lepton (on electric plane), its electrical charge (polarity above, below or on the neutral reality axis), its observed mass (proximity to Higgs ground state at position 6), its relative stability (balanced polarity relative to neutral reality axis), its predisposition to form matter (whether it intersects the reality axis) and pairing with its generation particle partner (positional symmetry between the pair).

With all the fermions logically placed along the figure-eight boundary of the base-12 prime waveform as a probability map, we were then able to derive the probability of the six upper particles from the product of their three quaternion power terms of *magnitude, position* and *spin*. The probabilities of the boson particles were similarly determined according to their appropriate placement within the geometry of the base-12 prime waveform and their interpretation as sector areas of the base-12 unit circle.

The resulting probabilities for the upper fermions and the bosons were then adjusted by a factor of 10^4 to convert them from the fundamental units of the prime waveform to our conventional units of $(MeV/c^2)^2$. Finally, the mass of each upper particle in MeV/c^2 was given by the square root of its probability.

Once the six upper particle masses were determined, the probabilities of their six lower particle partners followed directly from their equivalence with upper particle mass, times the spin factor of the lower particle. Lastly, the mass of each lower particle was given by the square root of its probability.

In summary, the base-12 prime vibration appears to provide an accurate and intuitive blueprint of the roles and relationships of all fundamental particles of nature. It clarifies and simplifies particle physics through a visual geometric framework of how everything fits together, a surprisingly simple and mathematically beautiful theory of everything. Not only does the base-12 prime theory validate the Standard Model and honor the incredible accomplishments of all who developed it, it provides new insights regarding the nature of reality. Most importantly, it makes a compelling case that the base-12 prime vibration is beneath it all and may be the elusive missing link which unifies relativity theory of the big with quantum mechanics of the tiny.

Relationship between W, Z, Higgs Boson and Top Quark as Probability Areas

Having demonstrated how the probabilities and masses of the W, Z and Higgs bosons may be derived from the base-12 prime waveform, a further geometric relationship emerges which helps clarify the roles they play and their connection to the heaviest particle, the top quark. This relationship becomes apparent when we view the base-12 cycle in terms of radial probability areas in the complex plane.

The first part of this relationship is how the probability of the Higgs boson is half that of the top quark. As the Higgs boson probability of $12^6/2$ is defined by the four-position wide 7-to-<u>11</u> probability plane at position 6, it can be visualized as a radius of length 1 rotating through a four-position sector of the base-12 unit circle shown shaded in the top left circle below. This sweeps through a 120-degree sector equal to a third of the area of the circle, where the entire circle represents 100% probability. As the top quark's probability of 12^6 is twice that of the Higgs boson, this is equivalent to that same radius of 1 sweeping through an eight-position sector or 240 degrees of the base-12 circle. As the top quark's placement within the base-12 prime waveform is at the fully exposed spin orientation of position 3, its eight-position probability sector would be as shown below by the shaded area of the top right circle.

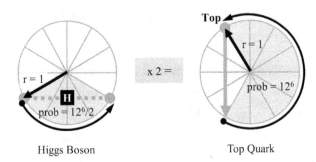

Higgs Boson   Top Quark

Higgs Boson and Top Quark as Probability Areas

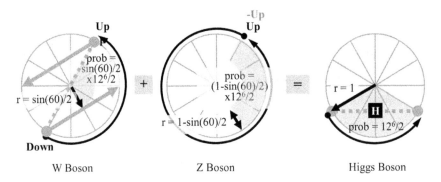

W Boson   Z Boson   Higgs Boson

W and Z Bosons as Probability Areas

So where the Higgs boson represents the horizontal and hidden ground state of the four-position wide 7-to-<u>11</u> probability plane of the Higgs (electromagnetic) field, the top quark is its vertical and fully manifested expression as a particle of matter. As an eight-position rotation in quaternion space is equivalent to a four-position rotation squared, this is none other than the defining relationship of $i^2 = -1$ between imaginary and real numbers in the complex plane. The Higgs ground state as i (unmaterialized probability below the reality axis) when squared (multiplied by i as a 90-degree rotation) manifests into the top quark as $i^2 = -1$ (fully materialized probability above the reality axis).

The second part of the probability relationship is between the Higgs, W and Z bosons in which the probability of the Higgs boson equals the sum of the probabilities of the W and Z. Assuming this equivalence is a logical consequence of probabilities as areas of the base-12 circle having to add to 100%, we can illustrate this as shown in the bottom three circles.

Viewing the top quark, W, Z and Higgs bosons in this way, as probability areas of the unit circle in the complex plane, further implies that all four of these particles have anti-particles. As an anti-top quark is known to exist there must be an anti-Higgs boson, albeit of the same zero charge as the Higgs boson. In turn, if an anti-Higgs boson exists there must be an anti-W^- boson (the W^+, known to exist) and an anti-Z^0 of the same zero charge as the Z^0.

I find it fascinating that the probabilities of the particles of nature can be expressed as simple areas of a circle with no need for advanced mathematics or irrational numbers such as Pi. This transports us back to the simple geometric ideals of ancient Greece. In fact, the original derivation of the area of a circle by the Greek mathematician Archimedes (287-212 BCE) was from a simple geometric relationship rather than a mathematical equation involving Pi.[55] The form of his equation was *Area = r x C/2* with Pi nowhere in sight. Pi was not even a number to the ancient Greeks but instead a ratio of two properties of a circle, circumference and diameter. Irrational numbers such as Pi = 3.14159... were philosophically offensive to the ancient Greek mathematicians as they didn't abide by the geometric simplicity of whole numbers and simple ratios such as that found in music. I guess that makes me kind of old-school too. OK, very.

This speaks to the ancient intuitive basis of so many of our modern mathematical concepts, a return to which would be a welcome simplification in my opinion. Still, many may say such a return to conceptual math would be an over-simplification with a detrimental loss of mathematical calculating power – just as many readers of this book may be thinking my base-12 prime theory is an extreme over-simplification of physics.

I would argue the counterpoint, that much of current mathematics and physics is an over-complication of what is needed to understand how reality works. I think the ancient Greeks had it right; that basic geometric shapes, proportions and relationships can fully describe the structure and beauty of reality and that whole numbers do indeed have numerological personalities which vibrationally express their energetic meanings.

Where advanced mathematics belongs I feel is in describing the complex phenomena which emerge from those basic geometric principles of nature. So both serve a critical role; one in understanding the fundamental rules and "intent" of nature and the other in examining the intricate complexity and creative diversity those fundamentals produce. It's just that modern science tends to assume complex approaches are needed for both and to underestimate the power of the simple as being too simplistic. Such is the double-edged sword of intellectual advancement; we often assume we need to pick up where we left off to make further progress. Sometimes the blackboard just needs to be wiped clean and for us to return to the basics to find a simpler truth.

The engineer and statistician in me would never have imagined something as baffling as the subatomic particles could possibly be explained by something as simple as the pattern prime numbers make in the base-12 cycle. It was only when I uncharacteristically ignored logic and followed my intuition that I was able to find the pattern at all. I also had the advantage of having little background in physics so that I could explore it from a new and naïve perspective, free of the technical rigor and scientific respectability of one who actually knows what he's talking about. So if ignorance is bliss, I'm one happy camper!

Seriously though, there is great value to bringing fresh eyes to problem solving as technical expertise in one area tends to lock us into a certain way of doing things. Sometimes it takes an outsider to help approach things from an unconventional, unvested and even simplistic angle.

The Emergence of Love at The Fundamental Scale

Before we wrap up our grand tour of the universe we call home, there's one further surprise the prime waveform has in store. And it's a tiny surprise of huge significance.

Recall that each fractal of the prime waveform takes shape once the required 10^{11} level of resolution is reached by the fundamental particles at that scale. To the observable universe those fundamental particles are its 10^{11} galaxies, to each average galaxy its 10^{11} stars and to each DNA molecule its 10^{11} atoms. However, we were able to take the prime waveform right down to the subatomic scale and accurately predict all the quantum particles and their properties by treating the atomic structure as a prime waveform fractal as well. What this implies is that the atomic fractal itself must also emerge from 10^{11} of an even more fundamental particle at an even smaller scale. This I suggest is the scale of consciousness itself and its particle expression, for lack of a better term, the *unit of consciousness*.

Although no such unit of consciousness has been detected as of yet, it's crucial to explaining everything else in our story here. In fact, it ties right back to the very start of this chapter where I first proposed that the prime waveform and the geometry of consciousness are one and the same – the perpetual flow of love consciously exploring itself. So although I presented the prime waveform from the cosmically large to the subatomically small, the unit of consciousness and the atomic fractal reveals that love actually emerges from smallest to largest – from simplest to complex, fundamental to composite, inner to outer. This leaves us with a stunning view of reality as nested fractals of manifestation all emerging from a creative spark of love beneath it all. But what shape would this enigmatic unit of consciousness hypothetically take?

As the base-12 prime waveform is a 3D double-helix structure with an inherent twist and directional flow, the simplest 3D geometry to describe those properties would be the tetrahedron – the simplest platonic solid. The four vertices or points of the tetrahedron makes it the simplest closed shape which connects any two positions of the double-helix probability boundaries of the waveform which, in turn, would make it the simplest building block from which the waveform could manifest. Upon inspection of the base-12 prime waveform we see that the geometric relationship between any two positions is indeed tetrahedral with four points of contact with the probability boundary. These four points of contact occur as two pairs of points, each pair defined by the orientation of the 7-to-11 probability plane connecting the two prime waves. This also accurately describes the polarized geometry of consciousness in its simplest form – as a point source experiencing itself outward in a dual polarized way.

The tetrahedron as the fundamental building block also makes sense in terms of it being the simplest shape that can express any direction or orientation within 3D and fully occupy that volume (unlike the even simpler sphere which leaves gaps when packed together). By varying its internal angles and the lengths of its edges, the tetrahedron is therefore the most versatile object for expressing the size, shape and direction of probability – both as an individual element and in combination. As such, the unit of consciousness could assume the shape of any type of tetrahedron according to its position within the prime waveform and vibrational geometry or energetic "intent" corresponding to that position.

This idea of the tetrahedron as the fundamental 3D geometric form of probability from which particles of matter emerge is not so far-fetched when we consider the latest findings in physics and mathematics. First, we can cite the amplituhedron recently discovered by Nima Arkani-Hamed and Jaroslav Trnka, the volume of which provides the scattering amplitude probability of particle interactions directly without need for the laborious "sum of histories" approach of Feynman diagrams and without reference to space-time.[56] Similarly, in mathematics, the Langlands Program of conjectures posed by Robert Langlands is revealing even broader connections between number theory, harmonic analysis and geometry.[57] As the base-12 prime vibration is a convergence of all three perhaps it is the underlying structure from which mathematics and physics ultimately derive.

Again, these developments indicate a growing trend in scientific inquiry back to the geometric ideals of ancient Greece – of fundamental shapes, simple proportions and whole number relationships. Recent advances in category theory further highlight this evolution of mathematics beyond rigid numerical equalities to the more universal relationship of *equivalence* – of how things are qualitatively related rather than quantitatively equal.[58] We will explore these vibrational qualities or numerological themes in depth in Chapter 5.

The everchanging fractal nature of consciousness and the reality it projects also reminds us why focusing on the present moment is so essential. It is our very latest experience in which we live that defines us, not the past or future. It is how we embody love *right now* that is our essence. Everything that came before was merely the experiential foundation to bring us to our current state of awareness and everything yet to be experienced are merely unmanifested potentials. Like a tree with its fractal legacy from seed to roots to trunk to branches to leaves, it is the current canopy of leaves upon which the entire tree's continued growth depends. Everything that came before, all previous seasons of prosperity or hardship, merely brought today's leaves to the elevation at which they happen to be. How those leaves seek illumination in this moment is all that matters.

The emergence of love from within to beyond also reminds us that it is how well we perceive, navigate and master love *within* that determines how well we can manifest love beyond ourselves. This not only means that we need to love ourselves unconditionally before we can unconditionally love others, but also that we first need to understand the language of love itself – consciousness.

This brings us to Chapter 4 in which we roll up our intuitive sleeves and learn how to engage consciousness through our emotions and extrasensory skills; natural abilities we all have but are simply rusty from lack of use. It is by fine-tuning our intuitive senses to the key of love that we can fully resonate with the prime vibration of All That Is and perceive its loving wisdom.

PART TWO

Becoming More Resonant with Love

4

Perceiving Love Through your Intuitive Senses

Intuition is that gut feeling or hunch we sometimes get about people and situations: a sudden vibe you feel when meeting someone for the first time, a friend calling you on the phone just as you were thinking of them or an urge to take a different route home only to discover you avoided a nasty accident. Despite how amazingly timely and helpful these little hunches can be, our logical mind prefers to write them off to more rational explanations such as reading body language, coincidence or just dumb luck rather than extra-sensory abilities of any sort.

Still, isn't it curious how certain people seem to get more than their fair share of these little hunches and do so in a particular way? You know, that guy in Sales who always seems to *hear* exactly what the customer wants before they even say it, the vet who is somehow able to *feel* where an animal hurts, the maintenance mechanic who just *knows* how to fix anything she comes across or the clothing designer with a knack for *seeing* what's going to be hot next year. These are all examples of individuals who are tapping into their intuitive strength, or what we can refer to as their *intuitive superpower*. Although such individuals seem a little more "plugged in" and naturally gifted than most, they are just normal folks who have simply embraced their intuitive superpowers and are applying those abilities in jobs and activities which make the most of them.

Whether we realize it or not, each and every one of us has an intuitive superpower, or combination of superpowers, which comes most naturally to us. As your intuition is a direct connection to your inner frequency of love, what you most love to do always points to what your intuitive strengths are. These are activities you are strongly drawn to and so thoroughly enjoy that, like the timeless state of position 6 itself in the prime waveform, time seems to stand still. This is that super-focused state of being in "the zone". And if you are following your passions, your career and hobbies align well with your intuitive gifts too.

Still, if you have yet to realize your intuitive superpower it's either because you're stifling it or you don't recognize it as a strength. This is very common as intuitive sensitivity is often mistaken for a weakness on the surface, especially in the business world and particularly with men who are often stereotypically expected to behave as unfeeling robots. I experienced that prejudice first-hand when I started embracing my intuition, and must admit I was also often guilty of it as the "old" me.

This anti-intuitive sentiment is very pervasive in modern society and influences how we feel about ourselves, and not in an empowering way. For example, an *empath* – one who deeply feels the energy and emotions of others – typically sees themselves as being overly sensitive and irrationally emotional. Until they embrace that sensitivity as a gift to fine-tune and utilize, an empath will tend to shy away from jobs and relationships which would expose them to emotionally charged situations – the very situations that need an empath's ability to care, emotionally relate and get to the heart of things. This I believe is the cause of much of the dysfunction and poor morale in businesses today. Many people are in the wrong roles, whether by their own choosing or by those doing the hiring and managing. They are consequently discouraged from developing their intuitive strengths to their fullest.

The good news is that our intuition never goes away, it's always patiently waiting in the background for us to connect. And even the most skeptical and resistant of us can do so at any time, we just need to be open to the possibility and get out of our own way. I should know – a former skeptic supreme who waited until he was 51 after enduring a 30-year manufacturing career which didn't really resonate with what he loves to do. I love exploring physical and spiritual truth and sharing what I learn with others. This is now what I pursue as a numerologist, medium and author and my intuition guides me every step of the way. And I couldn't be happier or more fulfilled. Time stands still for me most of the time now, although that balding guy in the mirror apparently didn't get the memo.

In this chapter you will learn how to recognize your intuitive superpower(s), develop those intuitive gifts and align your life with those callings. Remember, everything is consciousness in motion seeking to experience and understand love better. So understanding and enabling that natural flow of energy is ultimately what mastering intuition is all about. We are much more powerful and perceptive than we think and our intuition is the key to unlocking that potential for a truly inspired and purposeful life.

Your Extra-Sensory Abilities

Let's now meet your intuitive senses and get to know them better. Our five physical senses include touch, sight, hearing, smell and taste. We also have the intellectual sense of knowing. It is through these six senses that we perceive and interact with the world around us. Each of these senses also has a non-physical or extrasensory counterpart, named with the prefix "*clair-*" which means "clear":

- Clear Feeling = Clairsentience
- Clear Seeing = Clairvoyance
- Clear Hearing = Clairaudience
- Clear Smelling = Clairalience
- Clear Tasting = Clairgustance
- Clear Knowing = Claircognizance

For our empath, it is the extrasensory ability of clairsentience that enables them to feel beyond themselves. As the four senses of *feeling, seeing, hearing and knowing* come into play most at home and work, it is these four intuitive superpowers upon which we will focus.

To Feel: Following Your Emotional Guidance

"Don't be so emotional, keep your feelings out of it, you're being irrational". If you're an *empath* – one whose intuitive superpower is the ability to feel deeply – you have likely heard this type of criticism all too often. And, as an empath, you take such comments very personally. It has unfortunately become commonplace to label emotions and feelings as a bad thing, a distraction from the work at hand and a sign of weakness. However, as we will explore next, the exact opposite is true.

Our feelings and emotions are our primary intuitive guidance system, our most direct and tangible connection with our higher self and the universal energy field around us. Unlike the other three intuitive superpowers of extrasensory sight, hearing and knowing, which provide intuitive insights or "hits" in a more specific way, the range of sensations and feelings available through our emotions has a profoundly wide and diverse bandwidth. This makes the empath the intuitive superstar of them all in my opinion. Here we will examine exactly what an empath is, the classic traits of the empath and how those traits may be best utilized and nurtured.

So if you pride yourself on being a stone-cold unflappable poker face with no use for emotions, it's time to lighten up and reconsider the value of emotions to personal fulfillment. Take it from someone who's been there – keeping everything bottled up is a pressure cooker that will eventually blow off steam in unpredictable and unhealthy ways.

Feel me? Good.

Are You An Empath?

An empath is in touch with their feelings and the feelings of others. They relate well with people, have a natural desire to be understood and to understand others on an emotional level and have a keen sensitivity to the feelings, mood and morale of people and situations. This makes the empath an excellent judge of character. Yes, they are prone to outward expressions of emotion such as excitement or tears but that's a natural and healthy outlet for the empath to express particularly strong intuitive reactions and shouldn't be suppressed.

The empath has a knack for getting to the root cause of disagreements and personality conflicts because they can tune in to what others are feeling. They pick up on subtle energy cues and can often perceive the intensity of vibrational resonance or discord, from mild contentment or confusion to complete happiness or utter sadness. This also tends to make the empath a sponge for the emotions of others, taking external issues as their own and dwelling on matters long after the situation has passed. Quick to blame themselves for the behavior of others, empaths are an easy target for bullies and emotional manipulators. Whenever you see a chronically unhappy empath in your midst, you don't have to look far to find someone who is either intentionally or inadvertently taking advantage of the empath's sensitive nature. You may even find that certain someone staring back at you in the mirror.

Because an empath is so open to the emotional rollercoaster of others, they can feel overwhelmed in large groups and crowded public settings. In an attempt to avoid such sensory overload, an empath will typically shy away from group situations out of self-preservation and can appear introverted despite being highly social and engaging by nature. However, when feeling empowered and in their comfort zone, the empath is a delight to be around. They are able to focus intently on you, hold sincere interest in what you have to say and make you feel acknowledged and appreciated. Such is the amazing uplifting gift of the empath.

How To Empower The Empath

There's the misconception that you need to walk on eggshells around an empath for fear of hurting their fragile feelings. An empath's feelings are not fragile but simply highly tuned. They can easily detect avoidance and lack of authenticity which only reinforces their inherent insecurity about their sensitivity. So it's always best to be yourself around an empath and not try to handle them with kid gloves. Because they are so perceptive to the motives of others, treating them openly and honestly is how you gain the trust of the empath and empower them to be themselves too.

And once an empath feels safe, respected and empowered – Wow! They can be the most supportive friend, partner or family member and a company's greatest asset. Regardless what an organization happens to supply as a service or product, to run a business well is to manage relationships well. And the empath is the intuitive oracle of relationships. Able to quickly detect, engage and resolve small interpersonal issues before they become big, the empath is the early warning system and peacekeeper of the organization. More importantly, they bring heart and human connection to an otherwise mechanical and impersonal assemblage of procedures and processes.

As we will discuss shortly, the other three intuitive superpowers of extrasensory sight, hearing and knowing are most effective when also coupled with intuitive feeling. If, for example, you get an intuitive knowing not to step off the curb to cross the street, it's better if it's accompanied by a strong and urgent sense of danger to make sure you stop in your tracks. The same applies to intuitive business insights, again, because our emotions offer a broad range of feedback to enhance and emphasize otherwise subtle inner visions, sounds or thoughts. This is just one more way in which the empath is the intuitive superpower and nervous system of any energetically effective business or relationship.

The best types of roles for the empath is any capacity involving fewer people, a quiet and calm environment and which allows for personal attention, focus and helping others. Empaths often enjoy working from home or other such isolated settings because of their sensitive nature. They make great counsellors, therapists, health care/wellness providers, child care workers and personal coaches. Because they are so in tune with what appeals to others, Empaths are also well suited to advertising, marketing and branding. Regular opportunities to spend time outdoors in nature is very grounding and calming for the energetically conductive empath.

Examples of famous empaths include Jesus, Mother Teresa, Gandhi and Jane Goodall. Each of these individuals keenly felt the needs of others and channeled that sensitivity into serving the greater good in compassionate ways.

Roles not well suited to the empath include any job which exposes them to large groups, a hectic pace, volatile and unpredictable situations or frequent conflict. This would include managing many people, dealing with employee or customer complaints, legal activities, employee discipline or termination. It is also wise for the empath to minimize their exposure to social media and the inevitable insensitive comments by cyber ne'er do wells.

From the clear feeling of the empath we now explore the clear seeing of clairvoyance, of looking beneath the surface.

To See: Looking Beneath The Surface

Where the empath is the feeler of extra-sensory information, the clairvoyant is the seer. The strong visual intuition of the clairvoyant enables them to pick up on subtle imagery and translate abstract concepts in meaningful ways. This ability to clearly see the inner and outer worlds makes them the X-ray machine of psychic perception.

Are You Clairvoyant?

Do you have a vivid imagination and tend to daydream a lot? Are you good at seeing patterns, relationships and connections that others miss?

Do you like to surround yourself with beauty and art and does color strongly affect your mood? Are you a visual learner and need to "see it to believe it"? If so, you are likely clairvoyant.

The heightened visual acuity of the clairvoyant also tends to make them easily distracted, sometimes losing their train of thought. This also tends to make them easily distracted (just checking). They are also prone to reading too much into things as their imagination is so active.

How to Empower the Clairvoyant

Because clairvoyants prefer to perceive the world through pictures, they thrive in roles and environments which are visually stimulating and which challenge their creative vision. Clairvoyants make excellent artists, photographers, florists, decorators, designers, architects and engineers – basically any role which expresses pattern and form in aesthetically and functionally effective ways.

They have a keen sense of process and flow and are adept at simplifying systems and organizations so that all the pieces work well together. This ability to see the big picture makes the clairvoyant better at organizing and planning the broad strokes rather than executing fine details. Their distracted nature can make it difficult for them to focus on a specific task for any period of time or to pay attention when someone's talking. Wait, what?

Drawn to art, color and the beauty of nature from an early age, I am a typical example of a clairvoyant. I have always been a visual learner with a skeptical eye, needing to see things for myself. I had trouble paying attention in school and was a perpetual doodler, my notebooks being fantastically illegible at times. Arts, crafts and construction came easily to me leading me to a career in mechanical engineering. Roles which appealed to me most involved creating systems, improving process flow and streamlining operations – the typical big picture orientation of the clairvoyant. However, once the creative stage of devising a new system or process was done, I quickly became bored with the routine of day-to-day operations. This made me an enthusiastic systems implementer but uninspired operations guy to be honest.

In retrospect, my career choice of manufacturing was not a very good fit for me – an industry where predictable and stable production output is the name of the game and opportunities for creative expression few and far between. This no doubt contributed to me changing jobs and companies a total of seventeen times over the course of my thirty year stint as a plant guy.

As one's strongest physical sense is also typically their strongest non-physical sense, the first psychic ability to emerge with my spiritual awakening was clairvoyance. This included lucid dreams of highly memorable and vivid scenes, visual premonitions of things I would experience shortly thereafter and an inner sight of receiving pictures when I asked questions in my mind.

This inner sight began as indistinct purple and green shapes that constantly changed, sort of like a lava lamp (yes, I'm that old – Google it if you're a young whipper snapper). Over the course of weeks and months these shapes became sharper and clearer and took the form of recognizable images and objects, a visual dictionary of sorts. For example, a flower blooming became my symbol for "yes" while two hands crossing like an "X" became my symbol for "no". For the first couple of years, these messages were easiest to see with my eyes closed but eventually became perceivable with my eyes open too. I came to learn from other psychics and mediums that the development of this type of personal dictionary of symbolic imagery is the typical way that clairvoyance opens up.

Famous artists such as DaVinci, Michelangelo, Rembrandt and Monet were most definitely clairvoyant; apparent in their ability to see a finished work of art in a blank canvas or raw block of marble and then transform that clarity of vision into a masterpiece. Likewise, many of the top fashion designers such as Giorgio Armani, Coco Chanel, Oscar de la Renta and Vera Wang possess this gift of visualization and of foreseeing aesthetic trends before others.

To Hear: Listening To What Isn't Said

Clairaudience is the ability of extra-sensory hearing, of perceiving words and sounds through an inner voice. Unlike the mental chatter of our intentional thoughts which feel and sound like we are talking to ourselves, clairaudient messages seem to come from beyond. They have a gentle, calm and reassuring feel to them even when we are distressed or anxious.

I hear clairaudient messages in a voice that's different from my own; sometimes male, other times female and occasionally in the voice of someone in spirit I personally knew. In fact, my very first psychic experience back in 2014 was hearing my name "Michael" called out in my deceased brother's voice. I clearly remember it to this day and how different it felt from my own thoughts – it was as if the words came straight to my hearing rather than through my mind.

Are You Clairaudient?

Are you a good listener and able to pick up on what others are trying to say? Do you love music and the sounds of nature but tend to be woken up by the slightest noise? Do you catch yourself eavesdropping on other people's conversations? If so, I'm talking about you – but you probably knew that already.

Clairaudience often begins as a brief ringing or "tinnitus" in one or both ears and sometimes as a sensation of pressure fluctuations or muffled hearing. It then develops into hearing distinct sounds, words and even music and singing. I regularly hear bits of melody and lyrics in my mediumship readings when a person's loved one in spirit comes through with a tune that was meaningful to them both. I believe music comes through so often because it is very vibrational and thus easier for spiritual consciousness to vibrationally project. Electricity and light are also highly vibrational which is why flickering lights, calling cell phone numbers and turning appliances on and off is also very common though somewhat disconcerting at first!

Just as clairaudience usually develops from simple ringing, sounds or pressure drops to more complex perception of entire words, phrases and music, all extra-sensory abilities tend to develop in the same progressive way from simple to complex. As mentioned earlier, my clairvoyance or inner sight began as simple blobs of light before sharpening into distinct images.

That said, when there's an important message for spirit to give you, they will amplify it to make sure you receive it regardless of your level of psychic development. Such was the case with hearing my brother call out my name: although I was at the very start of my psychic development, I could hear it loud and clear. It was apparently time for me to awaken from the illusion of being separate and mortal to the eternal truth beyond the veil – quite literally my spiritual wake-up call.

How to Empower the Clairaudient

Roles which come naturally to the clairaudient include anything involving expressions of sound and voice. This includes careers in audio technology, music, singing and vocal training, teaching, public speaking, language translation, telecommunications, coaching and counselling. Essentially, any role involving the receiving, translation or delivery of audible messages is a good fit for the clairaudient.

They are adept at picking up subtle vocal cues that others miss, discerning what people really mean by the tone of their voice. This makes them good human lie detectors and also perceptive sales and marketing people, able to read potential customers well and tailor their pitch accordingly. Their auditory sensitivity easily overwhelmed by noise, a clairaudient needs quiet surroundings in order to focus.

Because they are keen listeners and communicators, clairaudients are prone to the extremes of either not speaking up for themselves or talking too much when they should be listening. They therefore go from passive and nonresponsive to chatty, animated and blurt out what they're thinking. Also, until they can distinguish higher guidance from their own mental chatter, they can have difficulty turning off their negative self talk and suffer from insomnia as a result.

Not surprising, a number of well known verbal channels were drawn to careers involving the clairaudient expressions of sound, music and/or communication. Lee Carroll was an audio engineer for 30 years before channelling Kryon. Darryl Anka, who channels Bashar, was first a special effects designer in Hollywood and then a writer, director and producer for his own movie production company. Wendy Kennedy, who channels the 9th Dimensional Pleiadian Collective amongst other entities and collectives, was initially in theatre and film. Micheila Sheldan, who channels many spiritual sources, was in marketing and communications before becoming a clairaudient channel of a variety of entities and collectives. So it seems those who are predisposed to the creative expression of sound and words in the physical world are natural communicators of channeled information too.

To Know: Trusting Your Inner Guidance

Unlike the more tangible perceptions of extra-sensory feeling, seeing and hearing, clear knowing or *claircognizance* is a sudden understanding of something you didn't know before. This often occurs as a flash of creative inspiration, a sudden insight or solution to a problem from out of the blue. Aha! moments such as these are direct downloads of higher guidance into your conscious mind. Because you can most easily access your intuition when in a receptive state of allowance, claircognizant messages often occur when you're relaxed and not thinking about anything in particular.

Are You Claircognizant?

Do you get a lot of creative ideas? Are you always curious how things work and why people do what they do? Do you have a knack for figuring out problems in your head and improvising quickly on your feet? Do you have a very strong conviction that your way is the best? If so, you are likely claircognizant. I wouldn't dare suggest otherwise as you're probably right.

Since intuitive knowing occurs as thoughts which bypass the senses, they can be difficult to differentiate from your own regular thoughts. This lack of sensory confirmation is why trust is so important for developing claircognizance; you need to trust the truth of these sudden insights and act on them accordingly. By acting on them, you give your other intuitive senses the opportunity to provide feedback – a gut check of their truth – whether to proceed or not. This is the fine art of using all of your intuitive abilities together as a skill set to guide your actions.

What you will likely find is that there is an "overflow" effect from your strongest psychic ability to your less developed ones. That is, the more you use your strongest ability, the rest will unfold too and start to catch up. This is very common with psychic development. A person's strongest skill emerges first and, if embraced and utilized, your other skills will be enhanced as well. Seasoned intuitives who have been open to their abilities for many years generally become quite well rounded in a variety of extra-sensory skills. This is how it is playing out for me too after seven years of doing intuitive work. My clairvoyant visions were first to surface, followed by claircognizant insights and then clairsentient feelings and clairaudient hearing. Clairvoyance is still my strongest sense but the others help supplement messages that come through.

How to Empower the Claircognizant

Roles well suited to the claircognizant are any activities which involve the generation of ideas, the acquisition of knowledge and innovation. This includes education, science and technology, research and development, writing and business. Because claircognizants can think and improvise quickly, they also make excellent entrepreneurs, problem solvers, first responders and even comedians – funny enough.

Quick witted and sure of themselves, claircognizants can be quite opinionated and strong-willed. They require a great deal of freedom and independence making them poorly suited to roles subject to a lot of bureaucracy, rules and protocol. Ironically, the fields most dedicated to the quest for higher truth such as education, science and social justice are also highly constrained by self-imposed tradition, conformance and protocol. Truly unconventional thinkers and innovators tend to be outliers to the establishment but rightfully so – they are the pioneers who break through barriers to truth and have the intuitive knowing to explore possibilities others won't risk. Those who I would consider to be claircognizant include Maya Angelou, Ellen DeGeneres, Amelia Earhart, Albert Einstein, Steve Jobs, Elon Musk, Nikola Tesla and Oprah Winfrey. The investment guru Warren Buffet similarly demonstrates an impressive intuitive knowing of financial markets, investment potential and timing.

Each of these individuals not only trusted and embraced their inner guidance but, more importantly, bravely acted on it using their given talents to the best of their ability. This brings us to Chapter 5 in which we explore the twelve vibrational themes through which our talents are expressed – the base-12 numerology cycle.

5

Exploring Love Through the Base-12 Numerology Cycle

In addition to having heightened intuitive senses to varying degrees, we each have natural strengths and talents which emerge in certain situations. I'm not talking about the skills and knowledge you acquired through schooling or job experience but rather those innate gifts and qualities which come naturally to you and make you the go-to person for specific types of activities.

Maybe you're a natural born leader with the unflinching self-confidence, drive and vision to take charge of situations and captain the ship. Perhaps you're the charismatic influencer with the communication flair, presence and power of persuasion to sell ideas and motivate. Or the perpetually kind and caring harmonizer with a knack for smoothing out conflict and bringing healing and calm to the most troubled of situations. These are but a few of the broad range of situational strengths you may possess, all of which fall under the twelve vibrational themes of the base-12 numerology cycle.

As mentioned in Chapter 2, the base-12 numerology cycle is not only the vibrational structure of the prime waveform but also the blueprint of consciousness which translates the waveform from probabilities into physical reality. As such, every conversation, relationship, career or life time represents a complete cycle of experience – a fractal of the prime waveform – in which we are turning potentials into reality. So how well we use our natural talents in handling each experience in a love-centered way determines how well the experience resonates or clashes with the prime waveform of love.

In this chapter, we examine the vibrational personality of the twelve themes of base-12 numerology and the unique situational skills they represent. Through the language of numerology you will learn the ideal *path of resonance* through the prime waveform and how you can best use your natural talents (and work around any weaknesses) to navigate that path in the most loving and authentic way. Also, as it is your intuition that links you to your true authentic self and the higher guidance of the collective consciousness, it is through your intuitive senses that you can know when and how to apply your talents for the greatest good.

The Twelve Themes of the Base-12 Numerology Cycle

To recap, numerology is the ancient science of the vibrational meaning of numbers where each single-digit number corresponds to a unique vibrational frequency with a specific meaning or "theme". Together, these number energies represent the universal cycle of themes we can encounter in any given situation from start to finish. In base-12 numerology, that cycle consists of the twelve number themes from 0 to 11 summarized below. Again, those themes at which we naturally excel are our born strengths, the types of situations in which we shine. Which themes best describe you?

   0. Potential – *The imaginative and empowering Enabler*
   1. New Beginnings – *The independent and confident Leader*
   2. Duality – *The cooperative and fair Balancer*
   3. Catalyst – *The creative and motivational Influencer*
   4. Structure – *The dependable and disciplined Organizer*
   5. Change – *The dynamic and adventurous Adapter*
   6. Love – *The caring and compassionate Harmonizer*
   7. Truth – *The understanding and wise Teacher*
   8. Manifestation – *The capable and productive Builder*

9. Completion – *The dedicated and thorough Finisher*
10. Awareness – *The perceptive and discerning Evaluator*
11. Illumination – *The inspirational and enlightened Illuminator*

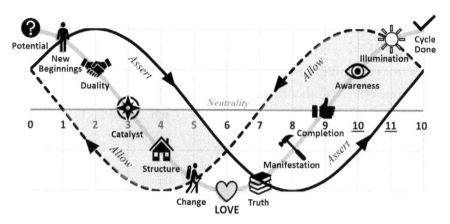

The Base-12 Numerology Cycle

The number 1, for example, represents the energy of *New Beginnings*. This includes the positive characteristics of independence, initiative and leadership – traits of the driven and confident leader. Yet, like any vibration, each number theme must oscillate up and down through both its positive and negative range to be fully expressed. Only by experiencing the full range and polarity of a number's theme can we fully understand and embody it. In the case of the 1 energy, its negative tendencies include isolation, aggression and fear. Thus, a person, team or company having mastered the 1 theme of *New Beginnings* excels not only at asserting independence, initiative and leadership but also at skillfully deterring isolation, aggression or fear from developing.

Given in the table below are the detailed meanings of the twelve numbers. Note how each number has both positive and negative attributes just as any vibration has an upper and lower amplitude, a crest and trough, through which it oscillates. And when I refer to positive and negative here I don't mean good versus bad but simply polarity as the full range of any number's vibration is necessary in understanding its true nature. It is this vibrational polarity and how it's positioned relative to the neutral reality axis that determines the specific probability geometry or proportion of each number, the window of potential between the upper and lower prime waves within which we can choose how to respond.

The geometry of each number's probability profile is also given to the right of the table to illustrate how the number's geometry precisely reflects its meaning. This includes its inherent spin, the one property of the prime waveform that is hidden from view in our physical perception of reality but nevertheless very real. It is what twists the singular wholeness of the base-12 circle into a dynamic double-helix spiral of duality in balance.

Spin is the angular momentum of the unseen, the depth beneath the surface and terrain within the map. In physics, as I posited earlier, spin is the hidden truth behind the illusions of dark matter, dark energy and the Big Bang. In daily life, it is the hidden momentum behind unexpected events and seemingly irrational behaviors and why some situations and people pull you closer while others push you away. Taking the time to understand the ever-changing spin behind the scenes helps us have greater patience and compassion for ourselves and others as we unravel the truth.

	Meaning	Positive Attributes	Negative Tendencies	Geometry In Prime Waveform
0	Potential	Optimistic, positive, enthusiastic, hopeful, open, empowered	Pessimistic, negative, apprehensive, closed, suspicious, hesitant, overwhelmed	Point probability, maximum upper polarity, outer limit of the waveform, fully backward spin
1	New Beginnings	Independent, original, innovative, confident, driven, accountable, leadership	Selfish, arrogant, aggressive, fearful, self-critical, isolated, vulnerable	1:0 ratio, mostly backward/partly downward spin
2	Duality	Cooperative, friendly, peaceful, inclusive, balanced, objective, fair	Competitive, critical, adversarial, biased, judgemental, moody, defensive	2:-1 ratio, mostly downward/partly backward spin
3	Catalyst	Influential, inspiring, motivational, social, creative, eloquent, supportive	Manipulative, opinionated, pushy, meddling, deceptive	1:-1 neutral ratio, fully downward spin, maximum probability range
4	Structure	Stable, practical, logical, realistic, organized, disciplined, dependable	Inflexible, stubborn, cautious, private, perfectionist, unrealistic	1:-2 ratio, mostly downward/partly forward spin
5	Change	Active, adventurous, adaptable, courageous, trusting, spontaneous	Restless, impatient, unpredictable, irresponsible, reckless	0:-1 ratio, mostly forward/partly downward spin
6	Love	Compassionate, caring, considerate, devoted, supportive, accepting, selfless	Self-sacrificing, doting, meddling, over-sensitive, needy, infatuated	Waveform origin and intersection point below reality axis, maximum lower polarity, fully forward spin
7	Truth	Knowledgeable, inquisitive, authentic, purposeful, spiritual	Pretentious, skeptical over-analytical, elitist, preachy, fanatical	0:1 ratio, mostly forward/partly upward spin

		Abundant,	Lacking,	
8	**Manifestation**	Abundant, productive, resourceful, capable, grateful, generous	Lacking, materialistic, entitled, ambitious, extravagant, wasteful	-1:2 ratio, mostly upward/partly forward spin
9	**Completion**	Fulfilled, content, accepting, forgiving, humanitarian, dedicated	Unsatisfied, defeated, vengeful, regretful, fatalistic, relentless	-1:1 neutral ratio, fully upward spin, maximum probability range
10	**Awareness**	Alert, present, attentive, focused, perceptive, intuitive	Distracted, confused, prying, paranoid, selective, obsessive	-2:1 ratio, mostly upward/partly backward spin
11	**Illumination**	Understanding, wise, insightful, receptive, knowing, enlightened,	Oblivious, unreceptive, naïve, disillusioned, ignorant	-1:0 ratio, mostly backward/partly upward spin

Table 1. The Meanings and Geometries of the Single-Digit Numbers in Base-12

As in the cycles of nature, the twelve numbers of the base-12 numerology cycle and their vibrational meanings follow a logical sequence and flow. This is the very same sequence we used to describe the life cycle of the universe, behaviour of galaxies, structure of the atom, manifestation cycle of the particles of matter and the creative process of consciousness itself. The base-12 numerology cycle is also universal in scope like the prime waveform upon which it is based. Every experience we encounter falls somewhere within this vibrational cycle of themes from the 0 of *potential* to the 11 of *illumination*. This applies not only to the small stuff of daily life but also the biggies such as relationships, family, health, career and our overall life path.

Also, just as the base-12 prime cycle passes through polarized quadrants of tangible reality (unshaded regions) and hidden potential (shaded regions), so too does the numerology cycle. However, as the numerology cycle applies to human free choice in which we can either *assert* or *allow* what shows up in our reality, these represent the two polarized approaches we can take. In other words, for every number theme we encounter we can choose how to respond; either fully asserting what we want, fully allowing things to unfold or somewhere in between.

If you ascribe to the idea of reincarnation as I do, that we are spirit (i.e. pure consciousness) having many physical life cycles, then you can appreciate how the base-12 numerology cycle also represents each life time we choose – from the 0 of conception to our return to spirit after 11. As such, each incarnation is our soul's conscious choice of returning to physical form to explore love from a different angle and through different circumstances, a unique opportunity to expand our soul's awareness of what love is and what it is not. And as the soul is a unique expression of All That Is/Source/God, every experience we have and choice we make enhances and expands love itself. So everything you do really does matter (and anti-matter).

Let's walk through the full cycle to clarify how all twelve themes work together.

0   Every experience starts from the 0 of *potential* and open possibility.

1   We step into that experience at 1 as a *new beginning*, feeling the independence, exposure and separation that creates.

2   As we engage the situation at 2, we experience the *duality* and contrast this new relationship brings.

3   From the polarized push and pull of the 2, we reach the *catalyst* of the 3 which urges us to take action one way or another to manage the duality.

4   From the drama of the catalytic 3 we seek the *structure*, stability and safety of the 4.

5   Eventually restless with the status quo of 4, we seek the freedom and *change* of the 5.

6   This gives us the courage and momentum to take a leap of faith towards the 6 of *love*.

7   Having experienced love, we are inspired and empowered to pursue *truth* and knowledge.

8 With knowledge and purpose comes the competence and will to *manifest abundance* of the 8 and, if spiritually motivated, selflessly for the greater good.

9 Our vision now manifested into reality, we reach the *completion* and acceptance of the 9.

10 From completion and acceptance comes the greater *awareness* of the insightful 10.

11 Through awareness we can see the lessons *illuminated* by the 11, bringing this greater understanding with us as we return to the 0 of *potential* to begin the next cycle.

Navigating Love along the Path of Resonance

Understanding the overall progression of the twelve themes we encounter in life as outlined above is supremely useful. It enables us to navigate any experience more proactively, confidently and calmly by knowing what next to expect.

Say, for example, you are interested in someone who just left a conflicted and unsupportive relationship in which they couldn't express themselves (i.e. the "2" theme of *duality*). That person will naturally seek a new relationship which resonates with the "3" theme of the *catalyst*, an empowering relationship in which they have a voice and can express themselves freely. If you know that you are not that person, that you tend to be demanding and want a partner who is more compliant, then you would both be better off with someone else. I realize this all just sounds like common sense, but we really do have a vibrational momentum pulling us from one theme to the next and helping ease us through the cycle in a self-balancing way. What we consider common sense is simply the wisdom of consciousness following the sensible prime cycle.

Still, is there a *best* way to deal with each particular theme? Again, to resonate fully with any theme's vibration means experiencing its full range of potential between the two prime waves. However, that's impossible as we can only choose one response in each moment. But the next best thing we *can* do is to choose a balanced response between the two prime waves, their net polarity when combined. That combined wave is the grey curve that runs through the center of the base-12 numerology cycle or what I refer to as the *path of resonance*. So to follow the path of the grey curve is to follow your most resonant path through the base-12 numerology cycle. This is the path that keeps you on track from one theme to the next while ensuring your overall approach is centered in love.

Let's revisit our cycle of twelve energies to understand what following the path of resonance actually involves. For each number theme I will describe what its probability geometry in the prime waveform represents and what the path of resonance suggests is our most balanced strategy for engaging that theme.

0 – The 0 has special meaning as emphasized throughout the book. Rather than a specific energy or frequency like the other eleven, it is a point probability which represents the beginning of each cycle and the unmanifested *potential* of all possible energies.

As indicated by the path of resonance being at its maximum positive amplitude at this position, the 0 theme is best approached with maximum positivity, optimism and enthusiasm. The direction of spin at position 0 is fully backward and thus begins the cycle from a perspective of initial detachment from the experience.

1 – The 1 is the energy of *new beginnings*. At its upper level (upper prime wave) the 1 is the assertive energy of independence, initiative, leadership and drive while at its lower level (lower prime wave), the hesitant energy of indifference, isolation and fear.

As the 1 has a +1:0 geometry with both the upper prime wave and path of resonance falling on the same point, our best strategy at 1 is to choose the highest, most confident response possible. This is in contrast to the lowest response being that of the neutral axis, of zero contribution whatsoever. Also, the spin at position 1 is mostly backward and partly downward. This amplifies the sense of separation and isolation which always accompanies stepping forward on our own.

As the 1 is the first energy we encounter in all new situations (birth, start of a new relationship, new job, new skill, new project, new day, new conversation, etc.) it determines the height of our initial trajectory into the entire cycle of experience and how strong a start we make to help carry us though the other energies. Thus, when beginning anything new, we should hold as positive and enthusiastic an attitude as possible.

2 – The 2 is the energy of *duality* or polarity. At its upper level it is the collaborative energy of encouraging cooperation, communication, diversity and balance and at its lower level the competitive and cautious energy of maintaining distance, discernment and boundaries.

The path of resonance at 2 indicates our most prudent course is not its highest expression of total cooperation as one might expect, but half-way between that and neutrality because of its +2:-1 geometry. This mostly positive bias of the 2 indicates that although we should always seek cooperative win-win outcomes, we also need to avoid giving away too much of own our personal power or suppressing our views for the sake of keeping the peace. A very important message here as it shows that speaking our truth and standing up for what we believe in, even if unpopular, is key to living an empowered life. The spin at position 2 is mostly downward and partly backward, urging us to maintain balance and stay grounded. Also, just as the trajectory with which we exited the 1 energy determined our vibrational entry into 2, how we handle 2 sets our angle of approach into 3.

3 – The 3 is the energy of the *catalyst*, that which causes change without changing itself. This makes the 3 an influential energy indeed. At its upper level, the 3 is the inspirational motivator and change-agent who makes things happen proactively while at its lower level the 3 just lets stuff happen and plays the victim.

The balanced +1:-1 geometry at position 3 places the path of resonance right on the neutral reality axis calling for unbiased objectivity and a balanced mix of proactive initiative and reactive readiness. We should therefore assess situations objectively in order to make an informed response rather than an emotional and rash one. As neutral objectivity can appear and feel vibrationally the same as not responding at all, it's easy to simply put your head in the sand and wait for the storm to pass. So, whenever you feel inaction is in order, do a quick gut-check to confirm if your inaction is well considered or simply avoidance.

Note too how position 3 (and its more enlightened counterpart, position 9) possesses the maximum probability range of the entire base-12 cycle and has a fully downward spin towards the reality axis. This is why the most massive top quark occurs at 3 (and the next heaviest bottom quark at 9) and why I referred earlier to the 3 as the incubator of diverse life. The 3 truly is the change agent of our vibrational reality. This makes catalytic situations very powerful opportunities to make course corrections in our lives when necessary. But the key is to see them as helpful detours guiding us to a better route rather than dead-ends spoiling the entire trip.

4 – The 4 is the energy of *structure*. At its upper level, the 4 resonates with the disciplined energy of stability, order and consistency and at its lower level with the more flexible energy of self-organization, informality and improvisation. Again, the 4 energy is our yearning for calm and order after the dynamic and often disruptive 3 of the *catalyst*.

The +1:-2 geometry at 4 puts the path of resonance half-way down below the neutral axis with a mostly downward/partly forward spin. This suggests less structure and formality while still ensuring sufficient order to maintain control and feel safe. This is consistent with the idea of the Law of Attraction (which we discuss in the next chapter) whereby we can manifest more effectively by detaching from specific outcomes and allowing the universe freedom to provide what is best rather than what we necessarily expect. A good reminder for all organizations too, that there should be a healthy balance between consistent operational standards and freedom to innovate and continuously improve.

5 – Where the 3 is the energy of causing change, the 5 is the energy of *change* itself. At its upper level, the 5 is all about action, movement and adventure – of seeking change. At its lower level, the 5 instead calls for adaptability, courage and trust – of allowing change.

The 0:-1 geometry at position 5 places the path of resonance all the way down to the lower prime wave which suggests our most beneficial response is that of adaptability, courage and trust. This means trusting our intuition to guide us in doing the right thing at the right time rather than impatiently scrambling to do many things at once. The upper prime wave passing through the neutral reality axis at 5 highlights how forcing change may feel very productive and practical but does nothing to help us change inside. Also, the mostly forward/partly downward spin at 5 gives us a satisfying sense of bringing things closer and more attainable when we embrace change.

6 – The 6 is the all-important energy of *love*. It is the origin and heart of the base-12 cycle and a point of timeless connection and unity where the two prime waves intersect. As such, there is no higher or lower level to unconditional love, only a state of harmony, compassion and joy that is already complete in and of itself. In contrast to the 0 of *potential* as the maximum unmanifested potential of the base-12 cycle, the 6 of *love* is its maximum potential realized. The fully forward spin of position 6 amplifies our sense of love as the most direct and powerful force we can experience.

Still, the path to love presents us with a paradox. For the two prime waves to merge at 6 and achieve love's full potential, they must meet half way below the neutral axis. This requires the physical courage at 5 to leap downward into the unknown but also the emotional courage to rise upward towards that connection. If we only pursue love in a physical sense of attraction and companionship, we can only realize a portion of love's full potential. We likewise only get halfway there if we yearn for love as a romantic ideal but are not willing to be vulnerable and take the necessary risk. We must completely allow, trust and surrender to love both physically and emotionally in order to fully comprehend it.

Also, note that it is at position 6 that the *downward* trajectory of the path of resonance from 0 to 6 reverses to an *upward* trajectory from 6 through 11. This is why love possesses such incredible power to uplift and enlighten, regardless how rocky the road may have been up until that point.

7 – The 7 is the energy of *truth* and the enlightened perspective we gain once we step through the doorway of love at 6. The upper level of the 7 falls directly on the reality axis and therefore resonates with practical knowledge and accepting how things are whereas its lower level resonates with pursuing your own truth.

The 0:1 geometry of position 7 indicates our path of resonance is that of the lower level, of finding our own truth rather than just accepting how things appear or automatically assuming the truth of others. The mostly forward/partly upward spin of 7 also helps instill a curiosity and sense of purpose for drawing truth towards us.

Note that the 7 of *truth* is the more enlightened counterpart to the 1 of *new beginnings*, their probability geometries being the same though opposite in polarity. This may be stated in a beautiful mathematical way as 1 + 6 = 7 or that the enlightened difference between 7 and 1 is the 6 of *love*. In the same way, all six positions from 6 to 11 are the *love inspired* higher frequency counterparts to the denser and more physically focused themes from 0 through 5: the 6 of *love* is the potential for love realized (0+6), the 7 of *truth* is a new beginning of love (1+6), the 8 of *manifestation* is love in physical/spiritual balance (2+6), the 9 of *completion* is the motivation of love (3+9), the 10 of awareness is the stability of love (4+6) and the 11 of *illumination* is being changed by love (5+6).

So where the 1 represented a *new beginning* calling for strength and independence of action, the 7 represents a *new beginning of belief* calling for strength of character and independence of thought. Also, where the path of resonance through the 1 put us on a downward trajectory for the first half of the base-12 cycle, the 7 begins an upward path of seeking greater truth and enlightenment.

8 – The 8 is the energy of *manifestation* and abundance, the more enlightened counterpart to the 2 of *duality*. Where the 2 explores duality and balance in relationships, the 8 explores physical/spiritual duality and achieving balance in our relationship with ourselves. At its upper level, the 8 is the energy of allowance and appreciation for what we already have while the lower level is the energy of manifesting abundance through will and effort.

The -1:2 geometry of position 8 places our path of resonance half way below the neutral axis, that is, more towards active manifestation than passive contentment. And just like the balanced figure-8 form of the number 8 itself, manifestation for the greater good requires a healthy balance between self fulfillment and providing for others. Indeed, the mostly upward/partly forward spin of 8 encourages us to be all we can be as the worthy and powerful manifesters we are.

9 – The 9 is the energy of *completion* and the more enlightened counterpart to the 3 of the *catalyst*. The upper level of the 9 is all about finding acceptance, contentment and closure when endings occur while its lower level represents dedication, commitment and finishing what we start.

Like the 3, the balanced -1:1 geometry of the 9 directs the path of resonance through the neutral reality axis. This calls for both the satisfaction of endings within our control and the mature acceptance of those beyond our control. Also like the 3, the 9 presents the greatest range of potential as to how we may respond. However, where the 3 prompts action to be taken the 9 involves gracefully stopping. The fully upward spin of position 9 also helps us appreciate the positive side of things, of the job well done or the life well lived.

10 – The 10 is the energy of *awareness*, insight and seeing things clearly. At its upper level, the 10 is the allowance energy of present moment awareness and intuitive openness whereas its lower level is the assertive energy of seeing what you want to see.

Like position 2, the -2:1 geometry of position <u>10</u> puts the path of resonance half-way above the neutral axis. This calls for a positive bias towards seeking the truth and seeing the good in situations and in people rather than making judgements and assumptions. The mostly upward/partly backward spin of <u>10</u> encourages this positive and honest outlook.

<u>**11**</u> – The <u>11</u> is the energy of *illumination*, of truths revealed and awareness expanded. The upper level of the <u>11</u> represents understanding a lesson being illuminated and adopting a new perspective whereas the lower level is that of remaining unchanged by the experience.

As at position 1, the -1:0 geometry at <u>11</u> places the path of resonance on the upper prime wave. This suggests our most beneficial approach to the <u>11</u> energy is to suspend what we think we know and be receptive to seeing things in a new light. Although these "got it" moments seem to occur when we least expect them they always reveal something we need to learn.

In terms of an entire lifetime of experience, the <u>11</u> is our final stage of conscious awareness in physicality before we return to the 0 of potential in spirit. As such, the <u>11</u> marks the culmination of all that we have learned while in physical form and the wisdom we carry back home to the collective consciousness when we physically pass. The mostly backward/partly upward spin of position <u>11</u> reflects this return back to the 0 of potential upon completion of the cycle of experience.

The path we have just taken I suggest is our most resonant route through the base-12 numerology cycle. It is the ideal path to be in synch with the love-centered prime vibration to which we are all tuned. Still, we will personally find certain energies more challenging than others and this will change as we gradually master various energies over the course of our lives and from one incarnation to the next. This is because we each choose specific themes to explore for each incarnation as our souls continue to grow. And to learn those themes, we intentionally choose life circumstances and personality predispositions (including a birth date and name) which draw us off the ideal path in those areas so that we have to navigate through that challenging terrain to find our way back. Your personal numerology reading reveals that plan.

Nevertheless, you can rest assured that our higher consciousness/soul always knows where we are on the board game. We just need to follow the voice of our intuition and where our passions lead as they will always guide us forward onto the most appropriate path – the path which enables us to learn what we came here to learn and to do so in the most fulfilling and love-centered way.

The Holy Grail of Love and the Incarnation Cycle

I know I've been tossing the word "love" around like a tie-dyed flower child who can't let Woodstock go, but I think by this point in the book you have come to appreciate that the principle of love is much more than just a romanticized concept or biological urge for something folks do while listening to Barry White songs. It is the fundamental essence of everything. Love is the vibrational signature of consciousness, the intent of reality and the purpose of our eternal spiritual journey.

Here's how I see things. When our consciousness is back home in its non-physical state between incarnations, we vibrationally reside at the home position of the prime waveform, position 6. Here we exist in a timeless state of love that is the inherent frequency and geometry of the 6. There we regain full awareness of our soul's true nature as conscious expressions of Love/Source/God. It is here that we know without a doubt we are creative aspects of love interconnected with everything else. We have a backstage pass to the full double-helix nature and 12D crystalline memory of the base-12 prime waveform rather than just its limited physical projection experienced by the audience. Once we remember who and why we are, it is our greatest joy to reincarnate again and again to expand our soul's understanding of love. And when our soul has evolved to a point at which we can better serve as guides for others, we may choose to remain in non-physical to expand love as part of the backstage crew.

When I first graphed the base-12 prime pattern and saw the perfect cup-shaped *path of resonance* formed when the two prime waves are combined, I confess something deeply resonated within me. I immediately thought of traditional depictions of the Holy Grail, the cup believed to be used by Jesus at the Last Supper and to collect his blood at the crucifixion. In that moment I had a deep sense that this Christian symbol is much more than a mere physical vessel and instead a universal metaphor for the cup-shaped potential of the love-centered essence of God, the same love-centered vibration of the prime waveform from which all emerges.

Perhaps this is why the idea of the Holy Grail has had such lasting appeal as a symbol of God's love in Jesus and Jesus' love in us. Maybe we have always intuitively known that the vibrational nature of God as love is in all things and resonates with this very shape – an energetic vessel made of love, to be filled with love and to be shared with others. The prime waveform reveals how this vibrational container of love is only possible through the combination of the two polarized prime waves – that our capacity to embody love requires the duality of both the physical and the spiritual.

Intentional Amnesia and the Illusion of Time

When experiencing the physical, we follow the path of the solid prime wave of matter and explore life as a linear progression of energetic themes from 0 to <u>11</u>. This not only gives us a false sense of *forward time* (left to right) but also polarizes our perspective onto primarily physical concerns; the two unshaded areas of probability in the upper left and lower right quadrants.

However, in preparation for our physical experience, we first temporarily supress our higher knowing that we are in fact eternal love in consciousness. This reversal of awareness or intentional amnesia occurs via the lower prime wave from 6 to 0, through the lower left quadrant of spiritual awareness. This essentially involves an initial lowering of our vibrational frequency to enable us to inhabit physical form. The side-effect of this lower frequency, however, is a fragmented view of reality. In the terms of quantum physics, reincarnation is essentially our human expression of the Heisenberg Uncertainty Principle which states that an object's particle-like position and wave-like momentum cannot both simultaneously be known with certainly. That is, when we experience ourselves in physical form as "particles" in a specific location and time we cannot fully experience ourselves as a nonlocal timeless waveform of consciousness in spirit.

Another helpful analogy is to picture consciousness like a strobe light of awareness; the less frequent the flashes of light, the less we perceive of the full continuous scene. Still, our rational mind splices these fragments together into a seamless movie convincing us that what we see is all there is to see. This is not unlike how I allegedly appear on the dance floor of any wedding reception I have been to. Although I am actually a graceful and elegant master of movement itself, all others apparently see is a chaotic, choppy and embarrassing display of awkwardness. Again, this is simply because of the inadequate disco lighting at such poorly equipped venues.

Choosing Our Role in the Play

Although our incarnation into physicality doesn't begin until 0, our perception passes through the reality axis at 1 on its way there. As this would provide a glimpse of physical potentials here, this is perhaps the point at which we vibrationally commit to the physical life and timeline we wish to experience for the upcoming life cycle.

This choice would be based on a range of criteria such as the culture, family dynamics, gender and physical traits we wish to explore this time around as well as any particular life path themes, experiences and personality traits associated with the numerology of a particular birthdate. Presumably, the numerology of our birth name is also decided here in cooperation with the higher selves of our soon-to-be parents.

A Physical Focus from 0 to 5

Our predominantly physical focus from 0 to 5 begins with the 0 of conception and the point at which the soul merges with our physical form. We then progress through the 1 of birth as an independent individual, the 2 of duality and relationships, the 3 of catalytic influences and creative expression, the 4 of structure and stability and the 5 of change and freedom.

Again, these are very much physically oriented themes which energetically draw us away from our sense of connection to each other or to any unifying divine source. This increasing sense of separation we feel from 0 to 5 is driven by the increasingly retreating and descending rotation (spin) of the upper prime wave throughout these stages. This is the same intensification of gravity we considered in our earlier discussion of galaxy formation and in debunking dark matter.

Not only does this induce a sense of separation but also directs our awareness downward and towards the lower potentials of the 1 to 5 themes represented by the dashed prime wave and the shaded area of probability of the lower left quadrant. So don't feel badly if you tend to enter new experiences with an initial sense of fear (1), resistance (2), disruption (3), insecurity (4) or restlessness (5) – we all do and are vibrationally meant to. It's simply the geometry of consciousness helping us experience the full range of each number theme from high to low.

Spiritual Awakening from 5 to 7

The progression from 5 to 7 is a particularly crucial phase in our conscious awakening, a transition from the physical focus on self to the spiritual truth of selflessness. However, here we are also presented with a gap of awareness in which the 6 of love remains hidden and elusive. This is the only stage within the prime waveform that dips below the reality axis of our perception – quite literally a "black hole" of awareness. Thus, to discover love requires a courageous leap of faith into the void – from the yearning for change at 5 and towards the powerful pull of love at 6. This is why the emotional abandon of love feels so profound, spiritual and like falling – our consciousness is vibrationally *falling in love*.

As revealed through the grey overall wave of the two prime waves when combined – the path of resonance – we see how position 6 expresses the full vibrational amplitude of 100% potential. This underscores how love is the fullest possible expression of our potential while in physical form, only matched by the love we experience at position 0 or 12 (10 in base-12) back home in spirit. In this sense, when we experience love we really are experiencing Heaven on Earth. It is also at 6 where the downward momentum of the path of resonance bottoms out and reverses to an upward trajectory from 7 through 11, all powered by the momentum of love.

From the 6 of love we then progress to the 7 of truth and knowledge. How fully we explore the physical/spiritual duality of the 7 frequency determines the level of truth we seek. The more focused we remain on physical aspects of truth, the more drawn we will be towards intellectual truth and practical knowledge. Likewise, the more focused we are on spiritual aspects of truth, the more we will seek spiritual truth and enlightenment.

Indeed, every position within the prime waveform presents this balancing act between physical and spiritual duality. Rest assured, where our focus is drawn between these two polarities at any given moment is nevertheless exactly where we vibrationally need to be.

Now that we are below the reality axis at 7 looking upwards, we can suddenly see the geometry of the waveform from 5 to 7 for what it truly is. No longer is the geometry of love at 6 shrouded beneath a mysterious void calling for caution. We now clearly recognize the 5 of change as a necessary stepping-off point to reach the polarity reversal of the 6 of love, the transformative merging of the two prime waves where the physical and spiritual become one in a still point of connection and wholeness. It is through this reversal of perspective at 6 that the 7 of spiritual truth becomes accessible at all. And the polarity reversal of 6 switches our perception upward in a dramatic, often life-changing way.

A Spiritual Focus from 7 to 11

The rest of our journey through physical/spiritual duality takes us from the 7 of truth and knowledge to the 8 of manifestation and abundance, the 9 of completion and acceptance, the 10 of awareness and insight and the 11 of illumination prior to our return home. Again, this second half of the human experience is inherently more spiritually focused and builds upon the more grounded frequencies which came before.

With our now upward and positive perspective courtesy of the 6 of love, the undulating drama of the physical lessons we experienced from 0 to 5 falls away into the distance while the upper prime wave of spirituality from 7 to 11 comes into sharper focus. This helps us appreciate all the trials and tribulations we went through from 0 to 5 in navigating the physical terrain of life to get us to a more spiritually appreciative place. And, in contrast to the receding and downward rotation of the solid prime wave of physicality from 0 to 5 and the increasing sense of separation and heaviness it evoked, it rotates instead towards us and upwards from 6 to 11. This brings with it a sense of increasing connection and closeness that a more spiritual outlook offers.

This doesn't mean the journey from 7 to 11 is necessarily an enlightened one. We have all seen in ourselves and/or others the know-it-all elitist (7), materialistic empire builder (8), uncommitted quitter (9), prying busy-body (10) and oblivious Ostrich with its head-in-the-sand (11). Still, this is simply consciousness exposing us to the full range of each vibrational theme so that we can explore them freely.

Another interesting dynamic of the figure-8 flow of consciousness is how the solid prime wave of physicality from 0 to 5 flows towards the 6 of love while that from 7 to 11 flows away from love. This explains why, in those portions of our journey before discovering love, we always feel like we are being pulled forward towards some greater meaning or sense of purpose. Then, once we know love, the journey from 7 to 11 always links us back to love and informs our choices and actions with greater compassion.

Landings and Take-Offs from the Runway of Life

There are two vibrational stages in the human experience during which we lose touch with reality and are totally dependent on others (except, of course, the time you did too many shots on your birthday). This is from 0 to 1 when we are in our mother's womb and again from 11 to 0 when we transition back to spirit. These probability regions, and the polarity reversal at 6, are the only stages of the prime waveform which are removed from the neutral reality axis. Thus, only through birth and physical death can we bridge these two gaps in awareness and complete each full base-12 cycle of life. Through the illusion of mortality we achieve eternity.

Also, the vertical separation between the upper and lower prime waves represents the gap of awareness between our physical and spiritual duality. We see that it is smallest immediately after we are conceived at 0, widens as we experience the 1 of perceiving our self as an independent individual and then continues to grow further apart through the terrible 2's of duality – of learning yes and no, right from wrong, do I eat the dog biscuit or not (true story).

This supports why young children seem particularly connected to spirit and can often sense the presence of loved ones who have passed. This is further aided by the brain wave frequencies of young children being naturally slower until the age of about 7, making it easier to connect with the inherently calm and meditative state of consciousness. After that point, the veil between us and spirit grows progressively thicker with the natural quickening of our brain waves and the experiential heaviness of physicality itself. This is compounded by family, schooling and society placing further restrictions on what we believe and how we should behave.

Love In The Balance

Another key vibrational property of consciousness is how the two prime waves work together to maintain the vibrational neutrality of the 6 of love. The polarized symmetry of the two halves of the figure-8 waveform ensures that each pair of symmetrically opposite energies vibrationally combine and cancel each other to restore neutrality. Specifically, the geometries of 1 and 11 cancel each other as do those of 2 and 10, 3 and 9, 4 and 8 and 5 and 7. As such, these symmetrically polarized pairings reflect the balancing act of consciousness; of neutrality achieved through the duality of equal and opposite polarities. This is how love explores itself without losing its neutral and non-judgemental sense of self – despite our reactionary tendency to "pick a side" of polarity and throw ourselves out of balance with love.

Still, it isn't until we experience enough of life under the illusion that it flows in a linear way from 0 to 11 that we can realize it actually flows from the 6 of love in a continuous self-referential way. This is the great truth that we seek to rediscover with each incarnation and why we agree to the temporary amnesia at the beginning of each life time. Love is the hidden treasure we seek to find within and each incarnation cycle a new treasure map to decipher.

This means that whenever we project the 1 energy of isolation, control and fear, for example, we automatically increase the potential of the 11 of illumination to help rebalance us back towards the neutral compassion of 6. This doesn't necessarily mean we will always recognize the imbalance being illuminated when it happens, but the opportunity to do so will be amplified.

This I feel is what the 9/11 terrorist attacks represented on a fundamental vibrational level. The collective consciousness of humanity was increasingly towards the negative aspects of the 1 energy of separation, control and fear to the point that it heightened the energetic potential of an extreme 11 event to profoundly illuminate and highlight that imbalance. By no means am I trying to justify terrorism or any acts of aggression but simply suggesting that we manifest the reality attracted by our beliefs, personally and collectively, to guide us back to a more loving state. The 9/11 attacks were a tragic display of brutality the likes of which I hope never occurs again. Nonetheless, this event revealed the vibrational dynamics of consciousness seeking to rebalance itself towards a peaceful state, urging humanity to seek out and eliminate the *causes* of fear and isolation before it manifests into terrorist ideology and violent extremism. The profound selflessness and compassion of the first responders who courageously rushed to the scene to help those in need is the very thing that illuminates the pointlessness of terrorism and attacks it at its source. So too was the immediate rebuilding of the twin towers into a single structure symbolizing unity and togetherness rather than duality and division.

Just as I believe that physical disease begins as a personal energetic imbalance so too do I see terrorism as a societal disease emerging from collective ideological imbalance. Babies aren't born as terrorists, they're molded that way by others with severely distorted values. Every child is therefore a new opportunity for society to break that mold. Although we may think we're powerless as individuals to change extremist ideology either here at home or on the other side of the globe, we are anything but powerless. We are all connected in consciousness and can change the world for the better with every compassionate and kind thought we hold and action we take.

In the same way the 1 and 11 themes work together to re-center our consciousness back to the neutrality of 6 so too does every other pair of symmetrically opposite energies: the polarized and conflicted 2 is balanced by the awareness and insight of the 10, the catalytic disruption of the 3 is balanced by the completion and acceptance of the 9, the insecurity and instability of the 4 is balanced by the self-worth and abundance of the 8 and the unpredictability and restlessness of the 5 is balanced by the spiritual knowing and purpose of the 7. And, of course, when we are already resonating with the 6 frequency of love itself, there is no rebalancing required at all. We are already aligned with the heart of All That Is.

This reminds us that humanity is vibrationally programmed to be fundamentally loving and good and that most will gravitate towards greater levels of awareness, connection and compassion. When extreme personal or global events do occur which are out of the peaceful flow of life, something is significantly out of synch with love in our personal and/or collective consciousness and is attempting to realign itself.

The Numerology of Nomenclature

As mentioned earlier, it is not only numbers such as birth dates which resonate according to the twelve vibrational themes of numerology but also words. That's because every letter in the alphabet resonates with its numbered position within the alphabet as illustrated below. So, in the same way that the various digits of a number are added together and reduced to a single digit to determine a number's vibrational meaning, the meaning of any word can be found by summing the position numbers of its various letters and reducing that number to a single digit.

0	1	2	3	4	5	6	7	8	9	10	11
Potential	New Beginnings	Duality	Catalyst	Structure	Change	LOVE	Truth	Manifestation	Completion	Awareness	Illumination
	A	B	C	D	E	F	G	H	I		
	J	K	L	M	N	O	P	Q	R		
	S	T	U	V	W	X	Y	Z			

Table 2. The Meanings of the Alphabet in Base-12 Numerology

The Vibrational Wisdom of Names

All of the words adopted over time by any language are not merely random strings of letters as convenient placeholders of meaning but were intuitively chosen based on their vibrational resonance. This applies to every word or name that enters our vernacular. This is also why the birth name we are given by our parents is actually intuitively chosen by our own higher self before we incarnate; it vibrationally resonates as our *Expression number* with how we can best express ourselves in a given life. Following is a sampling of the birth names of some famous individuals to illustrate how the numerology of their names accurately reflects how they could best express themselves.

Albert Einstein

Born "Albert Einstein" = 132592 59512595 = 1<u>10</u>/<u>11</u> + 35/8 = <u>11</u> + 8 = **17/8** which resonates with *Manifestation (8) through New Beginnings (1) of Truth (7)*. This describes perfectly his life's focus as a physicist exploring how the universe manifests (8) through new and original ideas (1) of how it all works (7).

Stephen Hawking

Born "Stephen William Hawking" = 1257855 5933914 8152957 = 29/<u>11</u> + 2<u>10</u>/10/1 + 31/4 = <u>11</u> + 1 + 4 = **14/5** which resonates with *Change (5) through New Beginnings (1) of Structure (4)*. This matches well Hawking's ability to create (1) stability and structure (4) out of unexpected change and exploring the unknown (5), both in terms of advancing our understanding of black holes and courageously dealing with his debilitating disease of ALS (where ALS = 1+3+1 = 5).

Synchronistically, Hawking passed on the anniversary of Galileo's passing and Einstein's birth.

Elon Musk

Born "Elon Reeve Musk" = 5365 95545 4312 = 17/8 + 24/6 + 10 = 8 + 6 + 10 = **20/2** which resonates with *Duality (2) through Duality (2) of Potential (0)*. Known for his ability as an effective (and sometimes conflicted) collaborator and team builder (2), he brings ideas and people together (2) to realize their collective potential (0). Steve Jobs also had a 20/2 Expression number and, like Musk, an affinity as a visionary imagineer who challenged the status quo in pursuing his dreams despite substantial setbacks.

"Lady Diana" Spencer

Born "Diana Frances Spencer" = 49151 6915351 1755359 = 18/9 + 26/8 + 211/11 = 9 + 8 +11 = **26/8** which resonates with *Manifestation (8) through Duality (2) of Love (6)*. This aligns well with her abundant and giving nature (8) and resolve to achieve that abundance despite her conflicted (2) pursuit of love (6).

Donald Trump

Born "Donald John Trump" = 465134 1685 29347 = 111/10/1 + 18/9 +21/3 = 1 + 9 + 3 = **11** which resonates with *Mastery of Self (11)*. This matches perfectly his singular focus on what he wants, unflappable self-confidence and driven need to control and lead on his own terms. He has truly patterned his life on *the art of the "deal"* (4513 = 1x12 + 1 = 11), a word vibrationally synonymous with his name and strategic style of negotiation.

Oprah Winfrey

Born "Orpah Gail Winfrey" = 69718 7193 5956957 = 27/9 + 18/9 + 310/11 = 9 + 9 + 11 = **27/9** which resonates with *Completion (9) through the Duality (2) of Truth (7)*. This reflects well her perseverance to finish what she starts and achieve humanitarian outcomes (9) and to do so through the exploration of physical/spiritual (2) knowledge and truth (7).

The Vibrational Wisdom of Words

If the prime vibration is indeed the creative signature of All That Is, I find it no coincidence that a term often used for God is the *Prime Creator* and that the core belief shared by so many cultures is that God *is* love itself. Thus, I find it no coincidence that in base-12 numerology the word "God" resonates with the 6 of *love* (i.e. God = 764 = 15/6) as does the ancient Hebrew name for God, "YHWH" (7858 = 2x12+4 = 2 + 4 = 6). Likewise, the term "All That Is" for everything manifest (8) through the physical/spiritual duality (2) of love (6) resonates with that very idea (All That Is = 133 2812 91 = 7 + 11 + 10 = 2x12+6 = 26/8).

As the order of the letters in the spelling of a word do not affect its overall vibrational meaning in numerology, the word "dog" as the mirror image of "God" also resonates with the 6 of love. Note too how this mirrored relationship between God and dog in terms of spelling is reflected in their conceptual meaning: where God is unconditional love itself, a dog is a reflection of that unconditional love in physical form.

Similarly, both Jesus and the Buddha (Sanskrit for "the enlightened one"), ascended masters who incarnated into physical lives to illuminate truth, resonate appropriately with the 11 of *illumination* (Jesus = 15131 = 11, Buddha = 234481 = 1x12+10 = 1 + 10 = 11).

Perhaps it's also no coincidence that it took me 51 years (51 = 4x12 + 3 = 4 + 3 = 7 of *truth*) to believe that such spiritual concepts were more than just religious platitudes, the hard-core skeptic I was. And how ironic that only by applying my engineer's skepticism and logic to examine numbers in terms of vibrations was I able to debunk my own skepticism about a creative Source. This I feel was the first stage of my life path (birth date) of the 10 of *awareness* – proving to myself that God is real and that we are all divine expressions of that same all-inclusive consciousness – and that sharing my journey with others is the next stage.

A further illustration of the intuitive way we choose words is revealed in those employed by science to describe reality. Although we consider words such as "prime", "quark" or "universe" simply convenient labels for describing physical things in the English language, they all have a deeper conceptual meaning derived from the twelve vibrational themes of the base-12 cycle.

Starting with consciousness as the underlying dynamic of that which defines the self, we find that the base-12 numerology of the word "consciousness" (3651396315511 = 4x12+1 = 41/5) resonates with that very idea of the *dynamic (5) of the structure (4) of self (1)*.

Next, the "base-12 prime vibration" (2115 12 + 79945 + 492912965 = 1x12+9 + 2x12+10 + 3x12+11 = 19/10 + 10/1 + 12/3 = 10 + 1 + 3 = 1x12+2 = 12/3) resonates with its precise meaning as the *catalyst (3) of new beginnings (1) of duality (2)*. Note how the word "vibration" itself also resonates as 12/3, revealing how the base-12 vibration represents a universal vibration. Likewise, the "universe" (35945915 = 35/8) as the ultimate fractal created by consciousness indeed resonates with *manifestation (8) of the catalyst (3) of consciousness (5)*.

A further numerological insight surfaces when comparing the "twelve" of the base-12 number system versus the "ten" of conventional base-10. Where the word "twelve" (255345 = 20/2) resonates with the full physical/spiritual perspective of *duality (2) through polarized cycles (2) of potential (0)*, the word "ten" (255 = 10/1) resonates with a more limited physical focus of *new beginnings (1) through individual cycles (1) of potential (0)*. So although the number "10" of *cycle completion* appears to be the same thing on the surface in both base-10 (1x10 + 0 = 10) and base-12 (1x12 + 0 = 10), only base-12 resonates fully with the underlying duality of reality and our true physical/spiritual nature.

This 10/1 focus on the physical alone is further echoed in those words science assigns to the fundamental aspects of the physical world: the "prime" (79945 = 2*10* = 10/1) numbers as the building blocks of all natural numbers, the "quark" (83192 = 1<u>11</u> = 10/1) as the building block of all matter, "matter" (412259 = 1<u>11</u> = 10/1) as the building block of all physical reality and "life" (3965 = 1<u>11</u> = 10/1) as the building block of all biology.

Note too how the "COVID-19 Coronavirus" (36494 19 36965149931 = 3x12+9 + 4x12+8 = 1x12+0 + 1x12+0 = 10/1 + 10/1 = 2) resonates with facing *duality (2)* but that its fundamental nature as a "virus" (49931 = 2x12+2 = 22) resonates with *mastering duality (22)*.

From embracing your intuitive senses in Chapter 4 and applying your natural talents intuitively through the base-12 numerology cycle in Chapter 5, we bring it all together in Chapter 6. Here you will learn how to work with the fundamental laws of vibration to live an intuitively connected, authentic and resonant life – a life which embodies the prime waveform and leverages the power of love in each moment.

6

Mastering Love Through the Universal Laws of Vibration

It's fitting we conclude our journey together with Chapter 6 – the frequency of *love*, the origin of the base-12 prime cycle and the truth at the heart of it all. Even the shape of the number 6 embodies its meaning: of a linear perspective spiraling inward to reveal the circular truth of eternity, the very path the base-12 cycle takes from the isolation of position 1 to the reconnection of position 6. Indeed, it is only at 6 that the cycle becomes apparent at all and that linear time and mortality are seen for the illusions they are.

Having explored why love takes the form of the elegant prime waveform (Chapter 2) and how it emerges into our reality (Chapter 3), we now learn how to best utilize our intuitive senses (Chapter 4) and natural abilities (Chapter 5) in mastering the fundamental laws of vibration through which love manifests here in Chapter 6. Much of what we will cover are not new metaphysical concepts by any means. They have been utilized for ages in countless ways and have stood the test of time because they work. What is new is that the prime waveform finally brings them all together under one unified framework, showing how they work seamlessly together as enablers of love.

The Six Universal Laws of Vibration

As mentioned in Chapter 1, there are six universal laws or fundamental properties of anything vibrational. These include *polarity, wavelength, frequency, amplitude, resonance and octave*.

Polarity is the most fundamental property of a vibration; its oscillation through both positive and negative in order for the vibration to exist. It is this polarity that defines the overall character of any vibration and its range of expression relative to neutrality. As it relates to the prime waveform as a vibration of consciousness, polarity represents the full range of experiences that are possible – what we perceive as contrast. Thus, the more polarity or contrast we experience the more fully we can understand a vibrational theme.

The **wavelength** of a vibration is the length of one complete cycle. Normally expressed in terms of linear distance, it can also be expressed in terms of any incremental measure of position. As the prime waveform is based upon the twelve whole number positions of the base-12 cycle from 0 to 11, its overall wavelength is 12. Every experience we encounter and choice we make therefore falls somewhere within this experiential cycle of twelve number themes. Only by understanding the full cycle can we know where we are within it and how best to move forward.

The **frequency** of an electromagnetic vibration such as that described by the prime waveform is equal to the speed of light divided by the wavelength of the vibration. As discussed on page 60, the frequency of the prime waveform is 1. Numerologically speaking then, the prime waveform resonates with an overall vibrational theme of *unity, wholeness and self (1)*. Each of the twelve number positions within that waveform are also unique incremental frequencies themselves, essentially a musical scale or octave of twelve ascending notes. As such, the prime waveform resonates both with an overall frequency of 1 and twelve individual frequencies from 0 to 11 which combine to achieve that overall unity. This means that each of the twelve frequencies or themes plays a unique role within each cycle of experience and that all twelve are necessary to an understanding of the complete experience. Thus the purpose of consciousness and our souls' progression is to explore all twelve themes as fully as possible.

A vibration's ***amplitude*** is the vertical magnitude of the vibration above or below the neutral axis, what we perceive as loudness or intensity in terms of sound. When we consider the prime waveform as a whole, its amplitude as a probability ranges from positive 0.5 (i.e. 6/12 = +50%) above the neutral axis to negative 0.5 below (-50%). Each position or frequency within the waveform, however, may have different upper and lower amplitudes which combine to impart a unique geometry or personality to the position's probability. Position 4, for example, has a positive amplitude of 0.25 (+25%) and a negative amplitude of 0.5 (-50%). This imparts a +1/-2 geometry to position 4, the vibrational energy of *structure and stability*, which calls for greater simplicity and flexibility rather than structure and order.

Resonance is the intensification of a vibration which occurs when another vibration of the same natural frequency occurs nearby. Even if one of the vibrations is at rest, it will begin to vibrate when the other matching vibration is active. This is the principle employed in tuning musical instruments to tuning forks of precise frequencies. This is also the principle behind the *law of attraction* – of our thoughts and beliefs vibrationally resonating with, or attracting, potentials of like frequency. In other words, like attracting like.

A higher ***octave*** of any frequency is a doubling of that frequency. The first octave is twice the original frequency, the second octave is twice the first and so forth. Because any octave is a harmonic multiple of its original frequency it sounds like a higher version of the same note, just as a high C in music sounds like a low C only at a higher pitch.

As we learned in the previous chapter, numerology describes octaves in terms of *master numbers* whereby each master number is a progressively higher octave. The first master number in base-12 numerology, the 11 of *mastery of self*, is a cycle or octave of 12 above the 1 frequency of *self*. Note how this returns the 11 to its original starting location and geometry of position 1 within the base-12 waveform which is why the 1 and 11 resonate harmonically with each other. This vibrational resonance applies between any double-digit master number and the single-digit frequency upon which it based. The 22 of *mastery of duality* is an octave higher still being twice the frequency of 11 (two cycles of 12 above the 2 of *duality*) while the 33 of *mastery of catalyst* is three times 11 (three cycles of 12 above the 3 of the *catalyst*), and so on. Each higher octave of a given frequency thus represents a higher level of understanding or mastery of the theme being vibrationally explored.

Let's now examine how these six vibrational properties may be employed individually and collectively in realizing our creative potential. And just as no two clouds, snowflakes or ocean waves are identical expressions of water so too are each of us unique expressions of consciousness exploring its various states of spirit, physicality and co-creative fluidity in between.

Polarity: Finding Clarity In Contrast

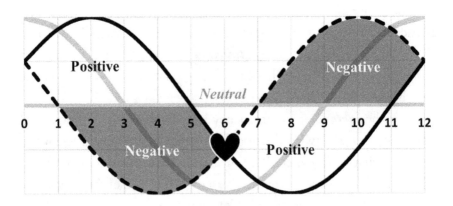

Polarity is simply the contrast between two conditions. It is through polarity that every vibration comes to life and any vibrational experience may be perceived. In terms of the geometry of the prime waveform and of consciousness itself, polarity is the 2:1 ratio of *wholeness through duality* and of *self-awareness through polarized experience*.

Neutrality Through Polarity

As polarity represents the range or contrast between two extremes, how fully we understand the vibrational range of any experience determines how informed and objective we can be. All of our choices and decisions are therefore driven by polarity and our ability to remain objective. Indeed, this principle of polarity applies to nearly every aspect of reality and takes many different forms: positive and negative, light and dark, feminine and masculine, open and closed, hot and cold, matter and anti-matter, physical and spiritual.

Although all polarity is valid contrast, our instinctively judgemental ego tends to play favorites and label things as good versus bad. This makes us quick to accept certain things and reject others, limiting how fully we can understand the whole truth. As mentioned in Chapter 1, our judgemental tendencies can lead to a physically biased scientific world view, a materialistic measure of success, spiritual skepticism and even extremist philosophies. To counteract such biases from limiting your experience of life, you should strive to face opportunities and challenges as objectively as possible as there can only be division and drama if you allow there to be. This means staying calm and being open minded to all possibilities and differing viewpoints before selecting your preferred course of action.

Now, this doesn't mean that your most appropriate response to every situation is necessarily a neutral one. As discussed in Chapter 5, each position or vibrational theme within the base-12 cycle has a net polarity which falls at a particular point on the grey curve, what we referred to as the *path of resonance*. In fact, it is only at the balanced +1:-1 geometry of positions 3 and 9 at which neutrality is the best response – of responding to *catalytic* events (3) objectively and accepting *endings* (9) graciously.

At positions 1, 5, 7 and <u>11</u> the polarity geometry is instead singularly polarized as +1:0 or 0:-1, calling for one polarity or the other – of embracing *new beginnings* (1) with full independence and confidence, *change* (5) with complete trust and courage, *truth* (7) with absolute integrity and honesty and *illumination* (11) with total acknowledgement and understanding. Conversely, positions 2, 4, 8 and 10 follow the lopsided +2:-1 or +1:-2 geometry of a 50% bias one way or the other – of responding to *duality* (2) in favor of cooperation and synergy over competition and separation, *structure* (4) in favor of flexibility and informality over order and predictability, *manifestation* (8) in favor of worthiness and pursuing your dreams over just accepting what comes your way, and *awareness* (10) in favor of being receptive and open to new ideas over asserting your own opinions. Last but not least, at positions 0, 6 and 12 the polarity geometry is doubled in favor of maximum amplitude either way – of maximum *potential possible* (0, 12) and maximum potential realized through *love* (6).

Although the polarity profile for each of the twelve vibrational themes follows the undulating geometry of the path of resonance above, below and through the neutral axis, the complete cycle of twelve themes achieves neutrality overall. This is the paradox of polarity: that overall vibrational neutrality is only possible through polarized experience. In other words, only by experiencing life fully can you truly be free of judgement, limitation and fear. Although neutrality can also be achieved by taking no action at all, by avoiding life experience entirely, that version of neutrality is not vibrational and therefore cannot resonate with the love-centered consciousness of who you really are.

Respecting Differences and Boundaries

Everyone explores polarity differently, at their own pace and in their own way. The ability to acknowledge these differences in others is an important part of mastering polarity yourself. We all need to learn and evolve on our own terms. And as we learn and grow, our perspective and boundaries are continuously shifting too. Boundaries are crucial as they define what *is* from what *is not*, what is within our control from what is beyond and what is acceptable from what is unacceptable. They establish our frame of reference and range of discretion for the choices we make. This includes limiting our choices so as not to cross the boundaries of others in an unwelcome or insincere way. As others often won't speak up when you have crossed the line, you need to be vigilant regarding your behavior.

Still, allowing others to choose their own path can be challenging at times, particularly if it involves someone close to you who seems to be jeopardizing their own happiness and potential. Even with those who seem to be making unwise and unloving choices, know they too are finding their way back to love albeit in a round-about way for a reason. It doesn't mean you need to follow or condone their choices, only to recognize it as part of their journey and to reaffirm what you prefer in contrast. It is only when we attempt to impose our views on others that we distort our own ability to see the truth. So it's up to each of us to honor our own truth without stepping on the truth of others. This is the key to mastering polarity.

Wavelength: Experiencing Full Cycles of Potential

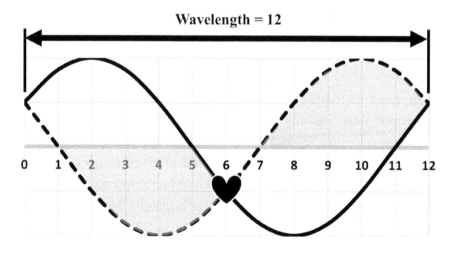

The wavelength of a vibration is its overall width or range of influence. It therefore defines one complete cycle. The wavelength of the prime waveform is 12 and spans a cycle of twelve incremental frequencies or themes from 0 to 11. From consciousness to atoms to galaxies to the cosmos, this is the base-12 cycle of experience through which everything is expressed.

Exploring The Wholeness of You

Where polarity is all about contrast, wavelength is all about completion and wholeness. To know the prime cycle fully is to know yourself completely and your relationship to All That Is. This is why every experience you encounter, regardless how small, is part of a cycle of learning and reconnection to yourself which falls somewhere within the twelve themes. Each theme relates to one another, building upon the theme before and preparing you for the next. We gained an appreciation of this sequential progression of themes in Chapter 5 whereby each number theme logically flowed into the next along the *path of resonance* – our path of least resistance to love.

Still, it can be difficult to see the big picture when we're caught up in the details and drama of one particular stage in the cycle. For instance, when experiencing some major upheaval in our lives – the *catalytic* 3 energy – such as the COVID-19 pandemic which threw us all for a loop, it was tough to see our way through to the 4 of *structure*, stability and safety. And even more difficult to see the light of the 11 of *illumination* at the end of the tunnel. That said, wasn't it amazing how the turmoil caused by the pandemic brought about a global sense of cooperation, community and "we're all in this together-ness" like never before, putting aside national interests and petty differences to work through the crisis together? Through shared hardship we rediscovered love in very profound way, recalibrating humanity back to what is important. How well we use that lesson in love to creating a more harmonized planet is now up to each and every one of us.

The pandemic removing our blinders and broadening our attention onto a shared challenge for all of humanity not only amplified a collective sense of community but also personal vulnerability, lack of freedom and fear. We saw numerous examples in the media and our own personal lives of those so self-absorbed to the point of denial, including not observing social distancing or the wearing of protective masks for the safety of themselves and others. This emphasizes how the universal vibrational properties work together in revealing the truth. Here, the pandemic illuminated the full *wavelength* of the global response and, in so doing, heightened the *polarity* of individual human nature within that vibrational response.

Uniquely Tuned to The Key of Love

Because the wavelength of the prime waveform is 12 its vibrational center is the 6 of *love*. This is the home frequency or key to which all other frequencies relate and are musically tuned. So whenever we authentically embrace our physical/spiritual truth, we naturally come from a place of love. Nevertheless, every physical cycle of experience we encounter necessarily begins from the 0 of *potential*, removed from the 6 of love. It is how we navigate back to 6 in rediscovering love that makes us physically complete and how we then apply love from 7 to 11 that makes us spiritually complete. And that journey back to love is as unique and diverse as we each are.

Love is always a personal voyage and an inside job. Although others can help us see what love is and what it is not, only can we discover and embrace it for ourselves. In this regard, we are like individual musicians of an orchestra playing different instruments but doing so together in the key of love. Individually we master our craft but only together does the full richness of love emerge into something greater that the parts. And this is precisely what the prime waveform represents – twelve unique vibrational personalities expressing their own unique geometry and polarized range, coming together as a complete musical scale from which anything is possible.

Frequency: Exploring Specific Vibrational Themes

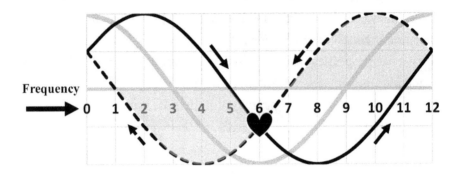

The frequency of a vibration is how many times its wavelength repeats in a given duration. As the base-12 prime waveform is a universal singular wavelength of 12 regardless of scale, it has an overall frequency of 1 as well as twelve individual frequencies from 0 to 11 as positions within the waveform. It is through these twelve frequencies which love explores and expands itself while resonating overall as one.

Mastering the Twelve Stages of Learning

In Chapter 5 we explored the twelve number themes of the base-12 numerology cycle and the vibrational geometry or "personality" of each. We considered how every experience we encounter may involve any or all of these themes and that those which come most naturally to us are our personal strengths or talents. How well we use those strengths to help us master the themes which do not come so easily is the higher purpose of every experience we encounter and every incarnation we choose. It is this dance of duality between the known and unknown – the mastery and mystery – which powers our growth.

We are all seeking the very same thing: to master all aspects of the human condition and to understand the full polarity of each of the twelve themes. In so doing, we evolve our soul's consciousness and expand the collective consciousness of All That Is. But how we choose to accomplish this is uniquely up to us as individuals as there are infinitely many different paths to mastery. Remember, you create your own reality and no one else can tell you how to do *you* best.

Also, enlightenment is not a race. Whatever you choose to explore in each experience or entire life time is entirely up to you, including how deeply you dwell on particular themes. Your goal is simply to follow the inner GPS of your intuitive senses and emotions (Chapter 4) to what resonates most and brings you joy. This automatically aligns you with your path of resonance through love. Yes, at times you will stray off the golf course and, other times, lose your ball in the woods. No matter – just chalk it up to experience, pull out another ball and focus on mastering your swing. This brings us to the importance of focus in achieving mastery.

Mastering the Moment

When we explore the full range or polarity of any theme we engage both the upper and lower prime waves. As the two waves always flow in opposite directions, engaging both effectively "freezes" time and motion and puts you in the timeless state of now. This is that wonderful Zen-like feeling of being "in the zone" when fully immersed in doing something you love. The only position within the entire cycle which inherently embodies that timeless state is at the intersection point of the 6 of *love*, so fully engaging a theme not only aligns you with the path of resonance to love but also resonates with the timelessness of love. This is why it's so important to be fully present in each moment and to love what you do.

Amplitude: The Amplifying Power of Emotions

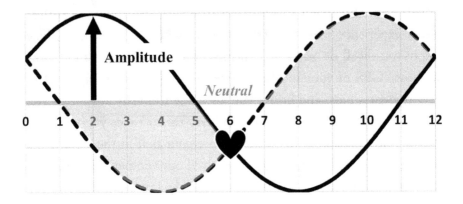

The amplitude of a vibration is its magnitude relative to neutrality. This is the characteristic which defines the power or emotional intensity of what is being expressed by the vibration. Every vibration has both an overall amplitude, which represents its maximum intensity, as well as local amplitudes at each position within the vibration.

Feeling Your Emotions Fully

In chapter 4 we explored how our senses, both physical and extrasensory, serve as our direct connection to Source. And the more fully we engage and utilize those senses, the more fully we can embody our physical/spiritual potential or amplitude. Still, it takes a strong sense of self to express our emotions freely, particularly in a modern society which praises steady conformance over emotional unpredictability. In my case, this didn't come easily.

Growing up in an emotionally reserved British/Canadian family, working in a technical career and being male all tended to suppress my emotions. It wasn't until my spiritual awakening that I finally confronted this limitation I was placing on myself and to see emotions as a superpower to be embraced rather than a weakness to be hidden. Interestingly, my numerology reveals that the 6 of love – of expressing love openly – is the one *karmic lesson* or past life weakness I carried over to this lifetime to work on deeply. I did this by intentionally excluding that number energy from my birth name: the letters of my name when converted to number energies gives Michael Peter Smith = 4938153 75259 14928 in which only the 6 is missing.

As our birth name or *Expression number* is how we choose to express ourselves, we will feel the absence of any number energy that's missing and have difficulty expressing that theme. So any theme you find particularly challenging is very likely a karmic lesson your soul wishes to master more fully this time around – and the only way to do that is to explore the full polarity of its presence and absence. In my case, that meant struggling through a career path lacking love and experiencing a reserved family dynamic to better understand and appreciate love once spiritually awakened to its true vibrational nature.

Emotions as Your Leverage of Love

Because the prime waveform is a single standing wave vibration, there is a direct relationship between the overall amplitude of its two prime waves and where those two waves intersect at position 6. For example, when the amplitude of all twelve themes are fully expressed as depicted above, the intersection point of position 6 is also at its maximum depth below the neutral axis. If, however, the overall amplitude were reduced by half, position 6 would likewise be only half as deep. In short, the more fully we explore the emotional potential of life the more deeply we can understand love.

Nevertheless, we each came into this lifetime as a work in progress, our personal vibration a somewhat dampened and distorted version of the perfect prime waveform. It is how we explore, expand and resolve those imbalances that we become more in tune with our true nature as Source energy. I believe that once we become vibrationally complete and have mastered the game of physical/spiritual duality we will no longer wish to reincarnate. We will then remain in non-physical as guides for others and for exploring higher frequency aspects of those non-physical dimensions.

Our soul's evolution and journey of learning is truly never complete just as there is always more to discover about love. Mirroring the infinity shaped cycle of the prime waveform and its unlimited toroidal paths of possibility, love is never complete but always becoming more.

Resonance: Manifesting with the Law of Attraction

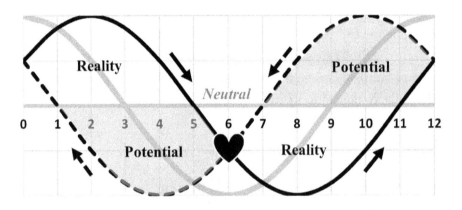

As mentioned in our discussions of polarity, frequency and amplitude, to know a theme well within the base-12 cycle is to experience and understand its full vibrational range. This enables us to find that point of net polarity along the grey combined wave, the *path of resonance*, at which we can best resonate with the prime waveform – our path of least resistance to love.

The "Classic" Law of Attraction

The law of attraction is the well known universal process of transforming thoughts into reality using the vibrational principle of resonance or "like attracts like". And this isn't only something we do to get what we *REALLY really want* (like a collector's edition *Spice Girls* lunch box) but actually do all the time whether we realize it or not.

As vibrational expressions of consciousness ourselves, we are always projecting both conscious thoughts and unconscious beliefs to the universe. However, the universe doesn't differentiate between wanted and unwanted beliefs. The vibrational thoughtform we hold, whatever it may be, is what the universe seeks to match with a reality of like vibration. Also, just as the twelve frequencies of the base-12 numerology cycle are general themes or archetypes rather than specific intentions, our thoughts and beliefs manifest in a general way unless we restrict them to a narrow outcome – something we all too often tend to do.

For instance, if you would love to visit Hawaii when you retire and can't stop thinking about the exhilarating freedom that tropical adventure would bring, it is actually the 5 theme of *change* that you are more broadly attracting. This therefore attracts a variety of experiences into your reality which resonate with that overall energy of change and freedom: maybe you are asked to travel to Mexico on an unexpected business trip, perhaps you run into an old friend who is a travel agent or suddenly you get invited to a destination wedding in Aruba, all expenses paid.

This highlights the key to effective manifesting: the universe, your Higher Self and your spirit guides know you far better than you do with your physically limited perspective. This is why it's always best to focus your thoughts on the *theme* you wish to attract rather than its specific form and to let the universe do the rest. If you hold out for a specific outcome, like that trip to Hawaii and *only* when you retire, you may miss many other opportunities and synchronicities in the meantime which would bring you the same type of joy in an even more timely and fulfilling way. So remember, the universe knows you best (please don't tell your mother I said that).

How the law of attraction works is basically through a three-step process which incorporates several of our six fundamental vibrational properties: of focusing your thoughts on a particular desire or *frequency*, increasing your emotional *amplitude* of that desire and then allowing the best outcome to manifest from the full *polarity* of possibilities. This three-step process of asking, allowing and receiving is how the law of attraction is traditionally described and the way I first learned about it.

Still, like myself, many have experienced mixed results in following this somewhat vague and simplistic formula. The great news is that the prime waveform clarifies and expands this process, explaining the specific vibrational principles involved in this powerful tool.

The Law of Attraction "New and Improved"

Actually involving six steps rather than three, the enhanced process of manifestation incorporates all twelve themes of the base-12 cycle in a vibrationally balanced way. To explain this, let's go back to what that cycle represents.

The base-12 numerology cycle is the creative blueprint of physical and spiritual duality from which everything emerges, where the first half-cycle from 0 to 5 represents physically focused themes and the second half from 6 to 11 their spiritually enlightened counterparts inspired by love. As such, each of the spiritual themes from 6 to 11 are vibrationally the result of adding the 6 of *love* to the physical themes from 0 to 5 (i.e. 0 + 6 = 6, 1 + 6 = 7, 2 + 6 = 8, 3 + 6 = 9, 4 + 6 = 10 and 5 + 6 = 11). These six pairings of 0/6, 1/7, 2/8, 3/9, 4/10 and 5/11 therefore represent the six steps involved when manifesting in a physically and spiritually balanced way.

Note too how the figure-8 path of base-12 cycle follows a forward direction along the solid wave from 0 to 12 and a backward direction along the dashed wave from 12 to 0. The solid wave represents the positive polarity of asserting a theme into reality – of projecting our thoughts forward – while the dashed wave is its opposite polarity of allowing potentials to happen – of attracting matching outcomes towards us. As each of the twelve number themes are defined by its probability profile between those two waves and their opposing energetic directions, it requires both assertion and allowance to "freeze" a theme in place long enough for it to manifest.

Let's walk through these six steps to understand how they work together as a vibrationally complete and love-inspired process of turning thoughts into reality:

Step 1: *Imagine a potential (0) you are passionate about (6)*
Step 2: *Focus (1) on that desire, believing (7) it is possible*
Step 3: *Emotionally relate (2) to the feeling of it achieved, knowing you are worthy (8)*
Step 4: *Allow synchronicity to guide you (3), trusting it will happen (9)*
Step 5: *Be flexible as to the form it may take (4), keeping open-minded and alert (10)*
Step 6: *Seize the opportunity (5) once it is illuminated (11)*

Here's a few additional tips to make the most of this manifestation process:

- In step 1, use your intuitive senses (see Chapter 4) to identify what truly excites you, not just what you think sounds exciting. We are all unique and each of us has a different picture of what would make us happy, so follow your own inner guidance rather than what appears to make others happy.
- In step 2, make sure you're focusing on what you want (e.g. a supportive relationship) rather than avoiding what you *don't* want (a controlling and insensitive partner).
- In step 3, it's important to hold the fulfilled emotion of already having achieved what you're trying to manifest rather than the unfulfilled desire of something you still currently lack.
- In step 4, remember to release expectations and allow the universe to bring to you what is in your greatest good without trying to force a particular outcome.
- In step 5, keep your eyes open for little signs and synchronicities that are leading you towards your new manifestation. Here, you need to go with the flow and follow your intuitive GPS for what feels right. And above all, be patient as impatience is just

another expression of lack and will undermine your best efforts to attract abundance.
- In step 6, have the courage to seize what comes your way. We all have regrets about opportunities we didn't act on, whether out of fear or complacency. So when positive things manifest, even if they aren't what you expect, grab hold and see where they lead.

Often when manifesting, your new reality comes together over a series of choreographed steps rather than one big payday. It's like playing chess; each move brings you closer to your goal but needs to be recalculated each step of the way depending on what the other player does. We are never creating our reality on our own but rather co-creating with the realities of everyone and everything affected by our choices.

Not to worry though – the universe has a handle on this intricate network of cause and effect, so trust the process and let the universe work for you.

Octave: Mastering Higher Frequencies

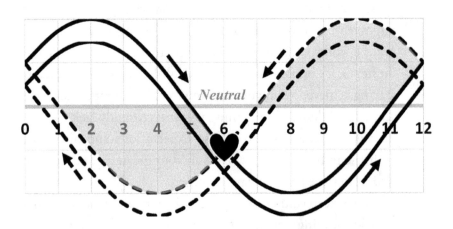

A higher octave of any frequency or musical note is a higher multiple of that note. The prime waveform as a cycle ending with a frequency of twelve represents the first such octave of the 6 frequency of love. Thus, it is both an octave of love overall as well as a twelve-note scale upon which theme specific octaves may be expressed. In geometric terms, the next octave of any number theme is one complete figure-8 lap of the prime waveform from that position back to that same position again – of re-experiencing a theme.

From Beginner to Student to Teacher

As with mastering any skill, practice makes perfect. Thus the more times we revisit a base-12 theme through different circumstances and context, the more we can master that theme in a comprehensive way. We can describe the process of mastery in general terms as occurring over three stages of awareness: from our first encounter with the theme as a beginner, to increasing familiarity and competence as a student and, finally, to deep understanding and mastering the ability to teach others.

As we learned in Chapter 5, higher octaves of the single-digit number themes are called *master numbers* in numerology. These are the double digit counterparts of the single digit numbers, such as 11, 22 and 33. Because each master number is the sum of the single-digit number being "mastered" (the second digit) plus that many higher octaves above (the first digit), a master number represents understanding a number both in terms of its individual frequency and in terms of complete cycles. Master number 33 of the *master catalyst*, for example, represents knowing the 3 theme of the *catalyst* personally, of dealing with unexpected events and influential change happening in your own life, as well as being a catalyst for others.

This emphasizes how mastery requires plenty of first hand experience in order to have the competence, knowledge and maturity to effectively teach others. Because teaching is a vibrational exchange between teacher and student, the more extensive the teacher's own life experience with the subject matter the better they resonate with a broad range of students. These are those special teachers and mentors we come across every once in a while who have a gift for meeting us at our own level and making learning enjoyable and illuminating. Just as all the full spectrum of colors is needed to create white light, a full octave of experience from 0 to 11 is needed to bring illumination and clarity. Thank you Mr. Saunders and Heather.

Because of the physical/spiritual duality of All That Is, the masters and teachers who guide us are not only individuals we encounter in physical form but also those in spirit who assist us from across the veil. This includes, first and foremost, our own higher self or soul as that greater aspect of ourselves which remains in higher consciousness, as well as individual and collective sources ranging from the souls of discarnate loved ones, spirit guides who agreed to assist us, ascended masters and angelic beings with whom we particularly resonate and even higher octaves of wisdom.

Whether we realize it or not, this higher guidance is always available. And as with any vibration, the guidance we can access is that with which we vibrationally resonate both in terms of our level of conscious awareness and the particular vibrational themes we are currently exploring. As such, the guidance we obtain changes and evolves throughout our lifetime as our consciousness expands and our journey progresses. I have experienced this myself throughout my spiritual awakening with the specific guidance and intuitive insights I have needed at each point of growth synchronistically appearing with perfect timing.

At the same time I first became drawn to base-12 and fascinated with the cycles of twelve in nature, I happened upon the website of Lee Carroll and his spiritual channels of the higher consciousness known as Kryon. These channels, beginning back in 1989, make numerous references to the universe being vibrationally structured in base-12 and emerging from a Source consciousness of love.[59]

Likewise, just before I discovered the base-12 prime vibration and realized it must emerge from a fundamental element of consciousness, I came across the channeling of the entity Seth by Jane Roberts dating back to the early 1960's.[60] Seth describes in detail the unit of consciousness as the fundamental particle from which everything emerges through electromagnetic manifestations, the very same conclusions reached by the prime waveform theory.

More recently, I became aware of Francesca Thoman's clairaudient channels of Nikola Tesla and the series of books they "co-wrote" together the first of which is particularly illuminating.[61] In these fascinating though quite technical channels, Tesla confirms the information of Kryon and Seth and dives much more deeply into the physics of consciousness. Again, all in timely alignment with my own journey.

Still, as often the case with channeled information, this guidance seemed quite cryptic and obscure at first and only made sense later once I had put enough pieces together on my own. Such is the way it seems with spiritual guidance; we only receive enough clues to assist us without doing all the work for us. After all, we wouldn't really learn if the answers were just given to us. In a way, spiritual guidance represents a higher octave of our current state of awareness which helps empower us to elevate our own vibrational understanding to higher and higher levels of mastery.

Octaves as Repeating Patterns

Much to the chagrin of our physically focused ego self, we sometimes find ourselves re-experiencing the same theme again and again. Perhaps it's adversarial bosses at work, needy relationships in your personal life or noisy neighbors wherever you live. This is often a tell-tale sign that there is something significant about this theme that you are not yet understanding. The whole reason we incarnate into physical experience in the first place is to learn what we *don't* know and become more vibrationally wise. Our higher self knows this and keeps placing onto our path opportunities to re-examine our chosen themes from a new perspective. As the old saying goes, the definition of insanity is doing the same thing over and over but expecting different results.

So when a theme presents itself as a repetitive challenge, sit quietly, focus on your heart center and ask for intuitive guidance as to why. Try to remove yourself from the frustration of the situation and see the repeating pattern as would a neutral observer watching a movie. How does the pattern fit into the story line and how does the lead actor (you) behave each time they face the theme? More importantly, how could the actor break the pattern to advance the plot in a positive and compelling way?

We really are eternal spiritual actors playing many finite physical roles. So when we get stuck on a particular theme, we are essentially type-casting ourselves into a certain comfort zone of acting. To escape that revolving door of limited roles we simply need to move beyond our comfort zone, act differently and demonstrate our versatility. After all, life should be more of an action adventure than a tired predictable drama.

Mastery of Play

Another key ingredient often overlooked in mastering our vibrational potential is that of play. Rather than a childish activity inappropriate for responsible adults, play is the most mature and spiritually evolved pursuit of all. It is the wave-particle somersault of imagination made real.

Play represents the ultimate freedom of expression, of stretching our imagination and seizing the infinite possibilities of the moment. When playing, we can be or do anything we wish and recapture the innocence, wonder and joy of childhood. Time becomes meaningless, the mundane magical and laughter effortless. Approaching life as a game simply to be played and enjoyed, not won or lost, brings an optimism and lightness to our actions. And even when things seems to take a turn for the worse, we know it's just a game and one in which we can imagine and improvise our way to a happier reality. As that's precisely how our higher self sees our physical journey, play allows us to gain that same joyous and engaged perspective of spirit.

In terms of base-12 numerology, play resonates with the 5 frequency of *change* and its liberated characteristics of freedom, curiosity and exploration. This includes the ability to adapt, improvise and travel which children so masterfully do in their minds and make-believe playground adventures. And as the 5 of *change* is our direct pathway to the 6 of *love*, a playful and enthusiastic attitude to life is so important to finding joy. This includes the improvisational playfulness of humor, of not taking yourself seriously and simply having fun.

It's not surprising that so many funny messages and comical images come through in my mediumship readings. Humor is one of the easiest vibrational exchanges across the veil because it resonates so effortlessly with love.

Although we all explore reality through the same six universal principles of vibration, how we personally do so is uniquely different. We are distinct expressions of love with the free will to choose our own way through every experience and lifetime. As our soul's awareness is the sum of countless diverse lifetimes, we are profoundly unique in terms of conscious experience and creative style. Tapping in to this vibrational uniqueness is the key to mastering play.

Still, embracing our creative originality within the confines of physical structure can be challenging. We are all subject to the same human biology, physical vulnerabilities and instinctive behaviors limiting how freely we engage reality. Our freedom of expression is also limited by societal, cultural and gender expectations and stereotypes of what we can or should do. We then further undermine ourselves with the falsehoods of mortality, that abundance is finite and that we have to compete with others to get our fair share. We basically live in an illusion trying to convince us of limitation and lack despite our unlimited creative potential.

This is why it is so empowering to explore your spiritual truth and to do so with improvisational freedom – insecurities and human frailties gradually fall away and your authenticity is liberated. The emotional and mental shackles once released, your spontaneity can take the stage with perfect timing and delivery. That's the liberating power of play.

The Collective Mastery of Soul Groups

As we explored in the previous two chapters, we each have particular intuitive strengths and practical talents we are continuously honing and mastering and various weaknesses we are working to overcome. However, no one person has complete mastery of all four extra-sensory skills nor all twelve situational skills, except perhaps the various ascended masters who have graced humanity's presence from time to time. For the rest of us, it's by sharing and combining our individual skills together through which collective mastery may be achieved. Indeed, I believe we seek a diverse mix of strengths and weaknesses in our soul families for this very reason – we complement, mentor and stretch each other through our contrasting differences.

As mentioned previously, the vibrational energy of your birth date – your Life Path number in numerology – is your main vibrational signature and the overall theme you personally came here to explore. Comparing the life paths of those in your family thus reveals how their various missions intertwine and serve as catalysts of growth and learning for each other. Likewise, adding those life paths together and reducing it to a single digit in base-12 reveals the collective life path theme you came here to explore as a team. I'll use my own family to illustrate.

My life path number is 28/<u>10</u>, my wife's 26/8, my mother's 16/7, my father's 17/8 and my five siblings 15/6, <u>11</u>, 13/4, 11 and 10/1. Combining all nine of our individual life paths together (using just the final single digit "outer vibration" of each) gives a collective family life path of 11 (i.e. <u>10</u> + 8 + 7 + 8 + 6 + <u>11</u> + 4 + 11 + 1 = 5x12 + 8 = 58 = 5+8 = 1x12 + 1 = 11). The 11 is the first master number or octave in numerology and represents the vibrational theme of mastering the 1 of *new beginnings and independence* or, simply, *mastery of self*. So although my personal focus is on a life path theme of the 10 of *awareness*, my family helps stretch my vibrational focus to achieve deeper mastery of the 1 theme of self, authenticity, independence and initiative.

Similarly, combining your life path number with that of your partner reveals how you vibrationally complement each other as a couple. For example, adding my life path of 10 to my wife's life path of 8 gives 10 + 8 = 1x12 + 6 = 16 = 1+6 = 16/7. Therefore, together, my wife and I enhance each other's ability to explore the 16/7 theme of *truth* (7) *through new beginnings* (1) *of love* (6). No wonder it was love at first sight for both of us!

As my Personality number (consonants of my birth name) is also 16/7, she helps me express my spiritual journey more openly with others. My Maturity number or focus for the latter half of my life (sum of birth date and birth name) also being 16/7 indicates my spiritual growth will continue to be enhanced by our connection. Further, as my Karmic Lesson (missing number energy from my birth name) is the 6 of *love*, my wife's love helps me overcome the very weakness I came here to address – again, by creating *new beginnings* (1) of *love* (6) together.

As you can see, our vibrational relationships with one another are just as important as our vibrational individuality apart, if not moreso.

Mastering the Spirituality of Science

As I shared earlier, what led me to venture into physics and culminated in this book didn't start out as a scientific endeavour at all nor did I pursue it in any scientifically rigorous way. I was simply following an intuitive trail of bread crumbs that began with a spiritually inspired obsession with patterns of twelve, prime numbers and ancient numerology. It was my trust and curiousity in pursuing these clues wherever they may lead – that is, by embracing the 5 of *change* – which uncovered the prime waveform and revealed the 6 of *love*. And only then did I realize I had happened upon a theory of physics, a deeper knowledge of the 7 of *truth*.

Granted, not your typical line of research. But is there really any right or wrong way to explore science as long as it provides learning, offers potential answers and stimulates interesting new questions? The trick is to view the discovery process not in terms of the accumulation of physical proof but rather as a never-ending deepening understanding of vibrational themes and the universal values they embody – that is, in terms of the base-12 prime cycle.

As we approach the end of the book, it's time to step back and revisit the traditional approach to science and challenge the longstanding materialist bias of the scientific method – of requiring physical proof for everything, of dissecting nature to its smallest parts and of keeping science and spirituality at arm's length. Once again we will call upon wave-particle duality as our gold standard for assessing spiritual-physical balance.

But first we need to come to grips with the nagging "measurement problem" of when something's a particle versus when it's a wave. Or, to borrow the Hershey's candy bar jingle, "Sometimes you feel like a nut, sometimes you don't". Also the motto I proudly live by.

The "Measurement Problem" of Science's Polarized Sunglasses

As discussed earlier, a quantum system behaves in a continuous wave-like state until measured or observed, at which point it seems to suddenly "collapse" into a fixed particle with a straight line trajectory. The century-old debate is still ongoing amongst physicists as to why this happens with a number of competing explanations for the phenomenon. The pilot-wave theory as proposed by David Bohm suggests that the hidden waveform guiding the particle trajectories never actually collapses when observed; it merely determines the location of the particle when we happen to look for it. Another feature of Bohm's theory is that both the observer and what is being observed become quantumly entangled together as their own combined wave function.

The prime waveform theory seems to support Bohm's conjecture when we consider the holographic fractal nature of the waveform suggested by our theory. Each fractal is only observable as a single distinct pattern or "particle" when viewed from a higher fractal perspective.

For example, the distinct platter-like shape of a spiral galaxy formed by its aggregate of billions of stars is only apparent when viewed from beyond. When viewed from within, however, such as our planetary vantage point within our own Milky Way galaxy, that cohesive singular form is undetectable. All we see is a seemingly random expanse of individual stars dispersed in every direction and behaving as separate particles. We essentially become an entangled observer within the galactic fractal oblivious to its wave-like nature. The same applies to the atom which appears as a single solid particle until we zoom in to the scale of the nucleus and realize it too is an aggregate of smaller particles, in this case electrons orbiting a nucleus. Although we can now perceive its subatomic constituents, we lose sight of its greater atomic form one fractal up the reality chain.

So when we attempt to observe which slit a photon of light passes through in the classic double-slit experiment and do so with a measurement method which can discern individual photons, we become entangled within the same fractal comprised of photons as particles. And because our observation is always "polarized" along a linear line of sight, that polarizes the photons we are observing too locking them into a linear path. That means no interference pattern on the screen downstream of the slits. Conversely, if we don't attempt to observe which slit each photon passes, we don't become entangled with them and therefore don't polarize their wave-like trajectories into linear paths. Thus, the interference pattern is free to occur.

Yes, all very speculative but what physics theory isn't? Besides, the fact that you've stuck with me this far means you must have some pretty speculative tendencies yourself. So, in a way, it's actually your fault that I've rambled on this long. Shame on you.

This leads us to speculate further on what would happen if we were actually able to observe all the potential photon paths at once rather than just one specific polarized outcome. Could this perhaps avoid the polarizing effect of physical observation we seem bound to in physical space-time? This would be like seeing from within the extra-dimensional vantage of 4D quaternion space, a space in which the base-12 prime waveform of all possible particle outcomes would come into view. This would include perceiving the toroidal flow of reality in the making and linear time as an illusion of lower dimensional physicality.

I believe this extra-dimensional perception is precisely what we tap into whenever we connect with our intuition. We are not sensing a set future that is cast in stone but rather the strongest potentials as they happen to stand at that particular moment. Free choice can always alter those potentials through new thoughts we make and actions we take but it's always the latest snapshot of what is in the *process* of manifesting that we pick up on.

Based on my own experience as a psychic medium, the messages I receive are a combination of past events which occurred with 100% certainty and potential future events with a good, but not certain, probability of occurring. Because of this, any psychic who claims to make 100% accurate predictions I believe is mistaken as absolute prediction is probabilistically impossible according to quantum mechanics and the continuously evolving nature of consciousness.

Exploring Nature Without a Hammer

Much has been learned about the structure of nature through essentially destructive methods such as particle colliders, biological dissection, radiation treatment and mining. Certainly, these have all proven instrumental in advancing scientific knowledge. However, this approach is what I would refer to as "brute force" science which breaks apart, bombards and excavates nature in order to examine the pieces. Unfortunately, what is left to pick up the pieces is inevitably nature herself.

Radiation and chemo therapy, for example, effectively destroy cancerous cells but are not selective; they destroy the healthy ones too and it's up to the body's own healing ability to recover from the collateral damage. Particle colliders such as the Large Hadron Collider at CERN can smash composite particles down to the level of quarks, leptons and bosons but the approach is more of a crime scene investigation than anything else. By examining the scatter diagrams left behind after a collision, physicists attempt to decipher the mayhem by identifying all the suspects based on how they fled from the scene.

Again, by no means am I downplaying the invaluable knowledge these approaches have facilitated as to the nature of reality and the technological advances made possible. Without the latest experimental findings of particle physics and subatomic structure would I have even realized that the prime waveform relates to the subatomic world at all. It's just that it seems we have pretty much broken everything down to its smallest pieces with nothing further to smash. I therefore believe the time has come to focus on non-invasive means of exploring nature in its natural habitat and understanding how it behaves at play instead of under duress. Astronomy is a perfect example and the Hubble telescope a star of that show. Likewise in health sciences such things as diet, exercise, rest, naturopathy, homeopathy, reflexology and energy healing all offer holistic and non-invasive means of encouraging the body's own healing power.

If, as the prime vibration theory suggests, everything is a manifestation of consciousness then everything is conscious and responds to conscious intent. So when we use destructive and invasive means of exploration I feel we project that aggressive vibrational intent into the very things we wish to observe in their healthy and whole state. Therefore, as long as our approach to science is to physically break things apart, we may ever only gain access to a fragmented view of reality because that is what resonates with our intent of fragmentation.

I suspect it will be for this same reason that any efforts to smash leptons or quarks into the even tinier hypothesized units of consciousness will likely fail, if it's even possible to build a collider with the vastly higher energy levels required. The idea of investigating love-based consciousness with destructive means seems counterintuitive anyway. Science should be an endeavor of both heart and mind to appreciate its full wave-particle nature, to embrace $E = mc^2$ as a beautiful relationship to be respected rather than split apart as a means to a sometimes explosive end.

Scientific Metaphors of Universal Truths

Metaphors enable us to facilitate understanding even further through analogy, by relating the commonplace and familiar with the new and obscure. Because metaphors are such an effective way to simplify and explain abstract concepts their use is widespread in educational and religious teachings. Classic examples include music as a metaphor for harmony versus dissonance, the spectrum of colours as wholeness through diversity, light and dark as clarity through contrast, numbers and planets as archetypal themes in numerology and astrology, the cycles of the seasons as a metaphor for the human experience and a wise and loving God as the personification of All That Is.

If the base-12 prime vibration is indeed the structure and intent of nature then all of these examples are more than merely abstract analogies. They are woven into the very fabric of reality as core properties of the creative process itself. I feel this is why so many of the metaphors we use to explain science and spirituality have had such lasting and universal appeal; they *are* the conscious intent of that which they describe.

Even what we consider as physical laws of physics may be better described as metaphorical physical expressions of deeper spiritual truths of consciousness. Newton's three laws of motion are a prime example: (1) that an object will remain at rest or in uniform motion unless acted upon by an external force, (2) that the force acting upon an object is equal to its mass times acceleration and (3) that for every action there is an equal and opposite reaction.

Restating these three laws of motion in terms of the underlying laws of consciousness from which the prime vibration suggests those phenomena emerge, we obtain: (1) that our physical reality and the patterns we attract will remain unchanged unless our conscious thoughts change, (2) that the power of our thoughts to manifest is equal to their significance to us and the emotional intensity we put behind them and (3) that for every inner thought we project we attract a matching outer reality. In this way, the physical phenomena of motion we observe in nature serve as valuable metaphorical hints into the nature of consciousness and the vibrational law of attraction through which it operates.

It is only when we attempt to distort such fundamental metaphorical truths into judgemental untruths such as an eye for an eye, good versus bad, worthy versus unworthy or loved versus unlovable that we distort their neutral benevolent meaning with our own human frailties and fears. The universe is non-judgemental and metaphors can be a powerful tool for reminding us of that wisdom.

Seeing the Sacred in the Geometry of Science

In the spirit of the prime waveform as an endlessly repeating cycle, I would like to end this book the same way it began; with the simple geometry of the base-12 circle.

Throughout our ambitious journey together we have considered how consciousness, the cosmos and all its apparent complexity may have emerged from the mathematical simplicity of the base-12 cycle and the inherent cyclical nature of prime numbers seeking to express themselves. Seemingly far too simple to be true, this conjecture only entered the realm of possibility by viewing the base-12 cycle as a vibrational standing wave of conscious experience endlessly repeating and re-expressing itself through polarity – that is, as consciousness experiencing itself through duality. And it was only when we interpreted the real number line as a circle, the circle as a cycle and the cycle as a vibration could we make that intuitive leap of faith. Only then could we consider that the vibrational geometry of *love* may indeed be at the heart of it all and the source of consciousness itself.

This suggests that love, at its most fundamental level, is the ultimate *why* behind every *what* we see in nature, physics and the human experience. It also reminds us that without the duality of appreciating what love is from what it is not can we vibrationally recognize it all. This realization has made a profound difference in my own world view. I now see all the ups and downs I have experienced– all the opportunities and challenges, the trials and tribulations – as the contrast necessary for exploring the vibration of love more deeply. This I am convinced is what we each came here to do in our own unique, empowered and compassionate way.

And as with any skill, only through trial and error and an unwavering desire to learn and grow can we master love more fully. It's so comforting to know that we always have a helping hand by the figure-8 geometry of love itself, continuously encouraging us *outwardly inward* to explore all twelve themes but always returning us safely back home to our origin of love at 6. So in a geometric, metaphorical and spiritual sense, love really is the point at the heart of it all.

Ironically, the search for truth through duality is the same approach taken by science already via the scientific method. Through the duality of theory versus observation and conjecture versus proof we assess what is "real" from what is "imagined" and, in so doing, expand our understanding of the real. As discussed earlier, this is how all particle masses are mathematically determined; from their probability waveform whereby squaring the probability as an imaginary value equals mass as a real number. And, again, this is because of the inherent cyclical nature of complex numbers in which an imaginary number (i) when squared ($i^2 = -1$) rotates it 90-degrees counter clockwise (negative) onto the real number line (-1). Note too how a further rotation of 180 degrees (-1 x i^2 = 1) then transforms that real number (-1 or matter) into its polarized opposite (1 or anti-matter). Thus, a full 360-degree rotation in the complex plane *must* manifest the duality of matter and anti-matter, the combination of which brings us full circle back to where we started (1 - 1 = 0) with the annihilation of both into the pure neutral potential from which they emerged.

Yet science tends to trivialize the fact that the discovery of the physically knowable is only made possible through an exploration of the unknown – that the measurable qualities of a particle are only knowable through its potentialities as a wave. All too often that originating structure of probability is treated as mathematical baggage and merely a computational means to an end once something physically measurable pops out of the equations. In other words, the unlimited potentials explored during the journey are less important than the specific physical destination reached.

Still, I look forward to the day when science and society acknowledges the fact that everything perceived as physically real are simply the most probable thoughtforms attracted into our field of view. And once that connection is widely embraced between consciousness and reality – between the sacred and the geometry – I feel we will advance our understanding and mastery of love in a dramatic way. There I go again, me and my feelings!

It's often said in metaphysical circles that the consciousness of humanity currently holds a vibrational frequency and dimensional acuity ranging somewhere between the third and fifth dimensions, with some who have heightened their consciousness to the sixth and beyond. With each such elevation in consciousness I believe we are able to more easily manifest thought into form although, paradoxically, we increasingly lose the materialistic desire to do so for ourselves in favor of the greater good. So the more we embrace the sacred geometry of who we are in body, mind and spirit the more we can recognize separateness as an illusion and that everything truly is connected through love, seeking love and is love.

NOTES

[1] A. Becker, *What Is Real?*, Basic Books, New York, NY (2018), pg.18

[2] Mission statement of Institute of Noetic Science (IONS), retrieved January 21, 2021, https://noetic.org/about/

[3] Mission statement of Foundational Questions Institute (FQXi), retrieved January 21, 2021, https://fqxi.org/

[4] Vision statement for John Templeton Foundation, retrieved January 21, 2021, https://www.templeton.org/about/vision-mission-impact

[5] Mission statement for John E. Fetzer Memorial Trust, retrieved January 21, 2021, https://www.fetzertrust.org/

[6] B. Wilson, *John E. Fetzer and the Quest for the New Age*, Wayne State University Press, Detroit, MI (2018)

[7] S. Post for J. Templeton, *"Is Ultimate Reality Unlimited Love?"*, Templeton Press, West Conshohocken, PA (2014), pg. xv

[8] Apple as 2020 World's Most Admired Company, Fortune, retrieved January 21, 2021, https://fortune.com/worlds-most-admired-companies/2020/apple/

[9] Apple as Forbes 2020 Most Valuable Brand, Forbes, retrieved January 21, 2021, https://www.forbes.com/the-worlds-most-valuable-brands/

[10] *Oxford Dictionary of Physics*, Oxford University Press – 8th edition, Oxford U.K. (2019), pg. 184

[11] M. du Sautoy, *The Music of the Primes*, HarperCollins, New York, NY (2004), pg. 84

[12] Millennium problems, *Clay Mathematics Institute*, retrieved January 12, 2021 from http://www.claymath.org/millennium-problems

[13] J. Currivan, *The Cosmic Hologram*, Inner Traditions, Rochester, VT (2017), pg. 95

[14] De Broglie–Bohm theory. *Wikipedia, The Free Encyclopedia.* January 12, 2021, https://en.wikipedia.org/w/index.php?title=De_Broglie%E2%80%93Bohm_theory&oldid=993602919

[15] G. 't Hooft, "Canonical Quantization of Gravitating Point Particles in 2+1 Dimensions", *Classical and Quantum Gravity* 10, no. 8 (1993): 1653. arXiv:gr-qc/9305008.

[16] J. Maldacena, "The Large N Limit of Superconformal Field Theories and Supergravity", (1997). arXiv:hep-th/9711200.

[17] *Oxford Dictionary of Physics*, Oxford University Press – 8th edition, Oxford U.K. (2019), pg. 600

[18] *Oxford Dictionary of Physics*, Oxford University Press – 8th edition, Oxford U.K. (2019), pg. 616

[19] M. Emoto, *The Hidden Messages in Water*, Atria Books, New York, NY (2001)

[20] Properties of water. *Wikipedia, The Free Encyclopedia.* January 12, 2021 fromhttps://en.wikipedia.org/w/index.php?title=Properties_of_water&oldid=999374831

[21] E. Siegel, *"Is The Universe Actually A Fractal?"*, Forbes.com (January 26, 2021) https://www.forbes.com/sites/startswithabang/2021/01/06/is-the-universe-actually-a-fractal/?

[22] S. Strogatz, *Infinite Powers*, Houghton Mifflin Harcourt, New York, NY (2019), pg. 48

[23] Complex number. *Wikipedia, The Free Encyclopedia.* Retrieved January 12, 2021, https://en.wikipedia.org/w/index.php?title=Complex_number&oldid=999297099

[24] C. Doran and A. Lasenby, *Geometric Algebra for Physicists*, Cambridge University Press, Cambridge, U.K. (2003), pg. 7-10

[25] "Dark Energy, Dark Matter", NASA, Retrieved January 21, 2021, https://science.nasa.gov/astrophysics/focus-areas/what-is-dark-energy

[26] J. Butterworth, *Most Wanted Particle*, The Experiment, New York, NY (2015)

[27] Equivalence principle. *Wikipedia, The Free Encyclopedia*. Retrieved Jan 12, 2021, https://en.wikipedia.org/w/index.php?title=Equivalence_principle&oldid=998220840

[28] "*The matter-antimatter asymmetry problem*", CERN (accessed January 21, 2021), https://home.cern/science/physics/matter-antimatter-asymmetry-problem

[29] S. Nadis, "*New Math Proves That a Special Kind of Space-Time Is Unstable*", Quanta Magazine (May 11, 2020) https://www.quantamagazine.org/black-holes-prove-that-anti-de-sitter-space-time-is-unstable-20200511/

[30] Cosmic microwave background. (2021, January 11). *Wikipedia, The Free Encyclopedia*. Retrieved January 12, 2021, from https://en.wikipedia.org/w/index.php?title=Cosmic_microwave_background&oldid=999663066

[31] E. Siegel, "*How Long Has The Universe Been Accelerating?*", Forbes (February 24, 2016) https://www.forbes.com/sites/startswithabang/2016/02/24/how-long-has-the-universe-been-accelerating/

[32] M. Starr, "*The universe is slowly dying*", CNET (August 10, 2015) https://www.cnet.com/news/the-universe-is-slowly-dying/

[33] *Where The Universe Came From*, New Scientist, Nicholas Brealey Publishing, Boston, MA (2017), pg. 93

[34] E. Battaner, E. Florido, "The egg-carton Universe", (1998). arXiv:astro-ph/9802009v1.

[35] M. Laychak, "Galactic census reveals origin of most extreme galaxies", Phys.org (September 11, 2020) https://phys.org/news/2020-09-galactic-census-reveals-extreme-galaxies.html

[36] *"Dark Matter Influences Supermassive Black Hole Growth"*, Harvard-Smithsonian Center for Astrophysics (February 19, 2015) https://scitechdaily.com/dark-matter-influences-supermassive-black-hole-growth/

[37] Laser Interferometer Gravitational-Wave Observatory (LIGO), Detection of a black hole binary merger, 2017. https://www.ligo.caltech.edu/news/ligo20171115

[38] Event Horizon Telescope collaboration. First image of a black hole, 2019. https://eventhorizontelescope.org/blog/first-ever-image-black-hole-published-event-horizon-telescope-collaboration

[39] K. Dickerson, *"Wireless Electricity? How the Tesla Coil Works"*, Live Science (July 10, 2014) https://www.livescience.com/46745-how-tesla-coil-works.html

[40] *"What Is DNA?"*, U.S. Department of Medicine (retrieved January 14, 2021) https://medlineplus.gov/genetics/understanding/basics/dna/

[41] B. Lipton, *The Biology of Belief*, Hay House, Carlsbad, CA (2005)

[42] G. Braden, *The Divine Matrix*, Hay House, Carlsbad, CA (2007)

[43] K. Cherry, *"How Many Neurons Are in the Brain?"*, Verywell Mind (April 10, 2020) https://www.verywellmind.com/how-many-neurons-are-in-the-brain-2794889

[44] N. Turok, *The Universe Within*, House of Anansi Press, Toronto, ON (2012), pg. 256

[45] Fine-structure constant. *Wikipedia, The Free Encyclopedia*. Retrieved January 12, 2021, from https://en.wikipedia.org/w/index.php?title=Fine-structure_constant&oldid=998556810

[46] M. Livio, *The Golden Ratio*, Broadway Books, New York, NY (2002)

[47] A. Vidyasagar, *"What Is Photosynthesis"*, Live Science (October 15, 2018) https://www.livescience.com/51720-photosynthesis.html

[48] K. Riesselmann, *"The Standard Model of Particle Physics"*, Symmetry Magazine (July 21, 2015) https://www.symmetrymagazine.org/article/july-2015/standard-model

[49] J. Currivan, *The Cosmic Hologram*, Inner Traditions, Rochester, VT (2017), pg. 9

[50] *"Long Sought Decay of Higgs Boson Observed"*, CERN (August 28, 2018) https://home.cern/news/press-release/physics/long-sought-decay-higgs-boson-observed

[51] Higgs field. (2020, December 6). *Wikipedia, The Free Encyclopedia.* Retrieved January 12, 2021 from https://simple.wikipedia.org/w/index.php?title=Higgs_field&oldid=7204272

[52] F. Wilczek, *A Beautiful* Question, Penguin Press, New York, NY (2015), pg. 305

[53] N. Turok, *The Universe Within,* House of Anansi Press, Toronto, ON (2012), pg. 76

[54] K. Riesselmann, *"The Standard Model of Particle Physics"*, Symmetry Magazine (July 21, 2015) https://www.symmetrymagazine.org/article/july-2015/standard-model

[55] S. Strogatz, *Infinite Powers*, Houghton Mifflin Harcourt, New York, NY (2019), pg. 33.

[56] N. Arkani-Hamed and J. Trnka, "The Amplituhedron", (2013). arXiv:hep-th/1312.2007.

[57] *"Modern Mathematics and the Langlands Program"*, Institute for Advanced Study online (Summer 2010) https://www.ias.edu/ideas/modern-mathematics-and-langlands-program

[58] *"With Category Theory, Mathematics Escapes From Equality"*, Quanta Magazine (October 10, 2019) https://www.quantamagazine.org/with-category-theory-mathematics-escapes-from-equality-20191010/

[59] L. Carroll, *Kryon Book 12: The Twelve Layers of DNA*, Platinum Publishing House, Sedona, AZ (2010)

[60] J. Roberts, *Seth Speaks*, Amber-Allen Publishing, San Rafael, CA (1972)

[61] F. Thoman, *Nikola Tesla Afterlife Comments On Paraphysical Concepts, Vol. 1,* Empowered Whole Being Press, Soquel, CA (2015)